# POMEGRANATE ROADS

## A Soviet Botanist's Exile from Eden

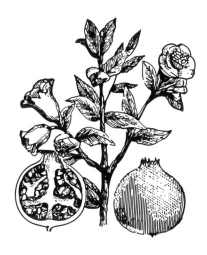

Dr. Gregory Moiseyevich Levin

# POMEGRANATE ROADS

## A Soviet Botanist's Exile from Eden

Dr. Gregory Moiseyevich Levin

Translated from Russian by Margaret Hopstein

**Floreant Press**

6195 Anderson Road
Forestville, California 95436
Copyright © 2006 by Floreant Press

Printed in the USA by Floreant Press
10 9 8 7 6 5 4 3 2 1
First Printing

Copyright Publisher's Cataloging-In-Publication Data

Levin, Gregory Moiseyevich.
  Pomegranate roads: a Soviet botanist's exile from Eden /
Gregory Moiseyevich Levin; translated from the Russian by
Margaret Hopstein; edited by Barbara L. Baer.


    p. : ill. ; cm.

Preassigned LCCN: 2006926681
  ISBN: 0-9649497-6-8

1. Botanists--Turkmenistan--Biography. 2. Pomegranate-
-Research--Turkmenistan—20th century. I. Hopstein,
Margaret. II. Baer, Barbara L. III. Title.

QK31.L48 P66 2006
580.92

Cover illustration by Daniela Naomi Molnar
Book design by Jody Grovier
Graphic production: Cyndi Reese and Windsor Green

In memory of my father, Moses Levin,
killed December 18, 1941,
on Pulkovo Heights defending
Leningrad from the German invaders

# CONTENTS

# Editor's Introduction

I wish I could explain why I got caught up searching for Dr. Gregory Levin and the wild pomegranates. My quest and failure to find either of them is the background story to *Pomegranate Roads.*

In the summer of 2001, while driving home on back roads in northern California, I heard words in Russian on Public Radio's BBC/The World. Anne-Marie Rueff, "The Savvy Traveler," was broadcasting from rural Turkmenistan, near Iran's northern border. "The birthplace of the pomegranate was here in the Kopet Dag Mountains of Central Asia. And here is one of the last places on earth where wild pomegranates grow."

Sonorous languages rose over the sounds of rustling leaves and cries of birds. I heard voices in Russian, English and Italian marveling at grapevines a dozen feet above their heads, at pungent arugula lying at their feet. "Wild pomegranates stretch their canopies along a riverbed," Anne-Marie said. This was Eden, a threatened Eden. The Russian speaker, a botanist named G. M. Levin, came back to the microphone. He said that conditions were going from bad to worse. "Here in wild pomegranate forests, sheep and cattle are grazing on grasses, eroding land, harming young trees." The persimmons, pears, apricots, apples, figs, and native grapes and his unmatched pomegranate collection—1,117 different varieties of pomegranates—at the nearby agricultural station of

Garrigala were dying from drought because the workers lacked pumps to bring up water from the Sumbar River. "We often carry water cans to each tree," Levin said.

I felt my throat constrict with thirst as I imagined Levin tending his parched orchards. In my mind's eye, he was a scientist from a Chekhov story, with a small beard and round glinting glasses. I supposed that he knew personally all the 1,117 pomegranates he had collected from twenty-seven countries on four continents.

Garrigala, the interviewer said, needed financial help. Since the breakup of the Soviet Union in 1991, Moscow had cut off funds to its former republics. From the Arctic seas to the Pacific Ocean, Soviet scientists had lost their labs, facilities and salaries—they were left out in the cold. To survive, many put their services up for sale. The need to re-employ nuclear specialists was obvious, but a pomegranate botanist less so.

Levin confessed that he was losing heart and ready to leave the station, but he lacked a successor and feared the collections would perish. Only help from the outside world could save Garrigala.

After the program ended, the fruit still glowed like rubies in my mind. I felt that more than chance had carried Levin's voice from Turkmenistan to my car radio. This man had been delivering a personal plea, an invitation for me to visit the wild pomegranates and do something to bring their plight to others' attention.

I first tasted a pomegranate early in life and never forgot my initiation into the mystery of the red globe. My mother wrapped me in a towel and gave me a spoon to explore the glistening treasure as only a child can—with fingers and seeds all over.

She introduced me to the Greek myth of Persephone. I saw myself as the dark-haired girl who wandered too far, while my mother was powerful Demeter who would never give up searching for me. My real mother with her shiny black hair and sun-browned arms transformed every apricot, peach and quince she grew into jams and chutneys. In August, the house always smelled of sugar, vinegar, ginger and cloves.

Pomegranates, each one an ovary packed with seeds, are the

essence of fecundity, femaleness and beauty. Some scholars now believe that Eve pulled down a forbidden pomegranate, not an apple, in the Garden of Eden. (This radical assertion will be repeated in *Pomegranate Roads*; Levin will also assure us that botanically speaking, Paris would have had to award Aphrodite a pomegranate for her beauty.) Persephone ate nothing in the Underworld until Hades, the dark Lord, tempted his captive queen into eating a few pomegranate seeds—sometimes the story says six, other times four. From her first taste, Persephone was doomed to return each year to the Underworld, causing Earth to suffer its winter death. The pomegranate is both temptation and a form of commitment and fidelity.

Persephone returns to Earth as the pomegranate trees begin to blossom, scarlet or peach or white flowers on branch tips, in the spring, After bees pollinate blossoms, flesh-wrapped seeds begin to grow around each original flower. As the abundant treasures expand within tough outer skin, the six-petaled calyx continues to hang down, revealing the fruit's plump sexual parts, as if the seduction were never over. In late fall, when most other fruit trees stand with bare branches against a grey sky, pomegranates blossom again and again, their bright amphora-shaped petals swishing like Persephone's skirts.

I was forced to postpone my quest to find Dr. Levin and the wild pomegranates of Turkmenistan for over a year. A month after I'd heard the radio program, September 11 happened. Nonetheless, I began applying for a visa to enter Turkmenistan even as American troops were setting up bases there to fight the war in neighboring Afghanistan. I also tried to contact Dr. G. M. Levin at Garrigala, but the research station had no internet connection. I finally found Dr. Muhabbat Turdieva, a plant geneticist at Tashkent University, who described the situation in Turkmenistan while I waited for my visa.

For a year, Muhabbat sent me news and forwarded Dr. Levin's articles that they had translated into awkward English. Levin, she told me, had arrived at the Garrigala Agricultural Station in the early 1960s and spent the next 40 years working in the subtropical oasis between desert and mountains. His station at Garrigala and the entire Sumbar River Valley with its mountain gorges and strange, subtropical flora were unique in the world. The more Muhabbat told me about this wild Eden, the more the visa to enter Turkmenistan was delayed, the more I was determined to go.

Early in 2002, I sent money to Muhabbat for a fund-raising brochure we created together in the hopes of reaching donors who would keep Garrigala alive. The small brochure was proof of my involvement, my intentions: "Ruby Treasure: Securing the Wealth of Pomegranates in Central Asia" felt like my own passport to Garrigala.

When my tourist visa finally arrived, I was granted a week in Turkmenistan, but only if I went with a tour group leaving October, 2002. Thus I found myself on a bus crossing the Amu Darya River, welcomed into Turkmenistan by a huge billboard-sized head and small squinty eyes of Turkmenbashi, the country's dictator. Whenever a village appeared in the stretches of desert, so did his little eyes, following us from above.

Saparmurat Niyazov, Turkmen Communist Party chairman when the Soviet Union collapsed, took power in 1991. Soon after, he renamed himself Turkmenbashi the Great, Father-Leader-Greatest of All Turkmen. In 1999, a rubber-stamp legislature elected him president for life. He proceeded to dedicate the resources of his impoverished country to his Stalinesque personality cult, renaming cities, streets, mosques, factories, airports and even days of the week after himself. And lest one forget him in the privacy of the home, his face appears on postage stamps, vodka bottles, tea bags and in digitally enhanced images on every TV screen. Beyond his borders, he's an international joke.

"You are fortunate to be in Turkmenistan in October," Ali, our guide, told me, "because October is the anniversary of the terrible earthquake, of our Independence Day, and the month our president published *Rukhnama*—our spiritual and practical guide to living." At that moment, I began to have a bad feeling about October. I knew from having lived in the former USSR that state holidays often brought out the worst paranoia. Officials kept a closer eye on foreigners, as if holidays might inspire sabotage.

Out of the desert blackness Ashgabat, the capital, glowed with a million lights. Massive pillared structures, inspired by the Greeks, Romans and Persians, stood eerily illuminated on the broad, empty avenues of this Cinecitta in the desert. Trevi-like fountains spouted plumes of colored water. At the highest point, visible for miles around, the golden statue of Turkmenbashi rotated 220 feet in the air.

"He always turns to the sun," said Ali, who had confessed he'd lost his job as an English teacher in a village school when Turkmenbashi eliminated foreign languages from the national curriculum. I remarked that the sun wasn't out. "He doesn't rest, not even at night," replied Ali. Wizard of Oz, I thought. Ozymandias.

The next morning I looked out the window of my Ashgabat hotel room through a hazy sky to the Kopet Dag range, which runs southeast along Turkmenistan's border with Iran and peters out near Afghanistan. The tan ridges, dangerous old crags on a major tectonic fault line, appeared deceptively soft through smog. Six stories below, hundreds of pigeons were using an Olympic-sized pool for a bath.

By the time my group left on a city tour, my email friend Muhabbat still hadn't registered. In her final message, she'd said she was having difficulties obtaining a visa from Uzbekistan to enter Turkmenistan.

Two modestly dressed people stepped through the revolving doors. One was a stocky Central Asian woman in a Soviet-era black suit with shoulder pads and a grim expression. I hoped she wasn't Muhabbat. Beside her stood an elegant slender man who looked like a Chinese sage.

The woman introduced herself as a native plant specialist sent by the government. She thanked me for my interest and then said, "Dr. G. M. Levin has recently immigrated to Israel. May I present the director of Garrigala Experimental Station, Dr. Makmud

Isar." Dr. Isar, she said, had taken a four a.m. bus from Garrigala to meet me.

"We are preparing for Independence Day," the woman said. "Because Garrigala is located close to the border with Iran, for security reasons the authorities have decided it is too dangerous for you to travel there."

"In three days, I'll be crossing into Iran myself," I said, my dream of visiting Garrigala vanishing before my eyes. "The holiday is two weeks away."

"Dr. Muhabbat Turdieva also failed to get a visa from Tashkent. Dr. Isar has come to show you the city." She wasn't apologizing nor backing down. I knew protest was hopeless. All directives must have come from the offices of Turkmenbashi.

Dr. Isar shyly presented me with two huge paper sacks of pomegranates. "We are so sorry for trouble," he said in Russian, looking embarrassed.

The pomegranates nestled like Christmas balls in tissue— garnet, cream-colored and hot-pink, tokens for all those fruit in a secret Eden I was not going to be seeing.

We approached the pseudo-Eiffel Tower on which the golden statue of Turkmenbashi slowly rotated. "He is always facing the sun," Dr. Isar said. As we rode an exterior elevator to the top, shabbily dressed people, open-mouthed and wide-eyed, pressed their noses to the glass. To these peasants, perhaps Turkmenbashi had built wonders. Perhaps he was a golden man, a khan of old with his raised arm.

"My daughters will prepare you Turkmen pilaf," Isar said. He hailed a cab. Almost immediately, the glass and concrete no-man's land of government palaces gave way to older apartments, then tenement walls and tilting wooden houses where laundry flapped on balconies in the desert wind. We stopped in a pot holed alley only ten minutes from Turkmenbashi's glittering stage set.

As the lamb for pilaf sizzled on a two-burner stove, Isar and his sons began to roll pomegranates out onto a flowered quilt. Isar expertly opened one after another. They were yellow pink, peach, crimson, maroon and purple—no two looked or tasted alike. Some had an acidic bite, others were boldly sweet, and a big pink one tasted like honey. As I swallowed the arils, my eyes filled with tears. I had not reached Garrigala after traveling so far, but I felt such sympathy for Isar and so welcomed by him that I could almost believe the forests stood before me in his living room.

"My pomegranate wine," Isar said, uncorking two murky bottles. "Good for digestion." The wine tasted medicinal yet sweet, like Génépi, the Artemesia-based liqueur from the Alps and Pyrénées.

After dinner, Isar lugged the pomegranates into a taxi and we drove back to Ashgabat's center. The night was warm, the imperial fountains cool. We wandered from one to another, sometimes talking, sometimes silent, reluctant to say good night. Arriving finally at the hotel, I thanked him. "For everything, the pomegranates especially," I said.

Two years after I'd heard Dr. Levin's voice, a year after my failed trip, something good happened. I contacted a dedicated woman named Shir Kamhi in the Agriculture Section of the Israeli Embassy in Washington. She treated my request as if it were a family matter, hunted among all the Levins in Israel for a pomegranate specialist until she found him.

The morning Gregory Levin's name popped up on e-mail, I was too moved to open it immediately. Until now, I knew him only as Dr. G. M. Levin. "Dear Madam Barbara I thank you for letter," he wrote, introducing me to the Web-based translation software that we would use for another two years. "It was very pleasant to found out that in U.S.A. there are people pomegranate interests."

As I knew, Gregory Levin had left Garrigala as the station was breaking up. The destruction of the valuable collections had been too painful to watch. His son and daughter-in-law had recently immigrated to Israel where he and wife, Emma Konstantinovna, joined them. He brought pomegranate cuttings with him and they were growing well in the hot dry climate. But he desperately wanted his life's work—what he, in Soviet terms, called a "theoretical book on pomegranates"—to reach an audience. He wished, "That work done has not gone to waste and is accessible to science. This is our common cause."

Again, Gregory Levin and his pomegranates became the focus of my attention. On a brilliant November day in 2004, I visited the Wolfskill Experimental Orchard at the University of California at Davis, a USDA National Clonal Germplasm Repository. Greenhouse manager Jeff Moersfelder and his assistant Joe Wehrheim spent their afternoon with me in the pomegranate orchards, muddy from the fall's first rains. They cracked open red, purple and yellow pomegranate globes for me to taste. Hanging alongside mature fruit, color-coordinated flowers glowed against the blue sky. The extended flowering time, Moersfelder explained, happened only in experimental orchards where numerous varieties grew together in what he called "a tree museum." In commercial agriculture, such as Paramount Farms' vast Pom Wonderful orchards in California's Central Valley, timing is not left to nature. There, all the thousands of acres of cloned Wonderful pomegranate trees bloom at the same time and produce fruit conveniently on schedule for workers to pick.

Moersfelder saved the best for last, walking me to a section of pomegranates marked PROVENANCE TURKMENISTAN. Parfyanka's fruit hung in abundance, each globe big and garnet outside, soft-seeded and wine-flavored within. Azadi nestled peach-colored in leaves. Girkanski stood out dark, almost purple, and when we tasted the arils, I drank in the sweet/tart flavor. Moersfelder said he wished they knew more about their Central Asian varieties.

"They came from Gregory Levin," I said, "from the largest pomegranate collection in the world in Turkmenistan." I then told them my failed quest story and how they could reach Levin in Israel. Days later, Gregory wrote about the UC Davis botanists. "Of course I reply I personally sent pomegranate varieties."

I'd been mulling over the idea that a book about pomegranate botany mixed with the story of Gregory's life and expeditions in the Trans-Caucasus and Central Asia might work for my small press. When I proposed this, he responded immediately. "Forty years I am engaged as hunter behind plants, gathering and creating collection. I am calmed by news and hope work is not vain, but also sometime be read. For this, gratitude does not have borders."

Gregory Levin's memoir with pomegranates is a survivor's tale, a botanical adventure that chronicles treks into regions far off most maps. Some chapters will especially interest botanists and pomegranate growers, while other readers will be led across mountainous Central Asia and the Trans-Caucasus. The subtropical Garrigala station and the wild pomegranate gorges he explored were not only home but a wild paradise that Gregory Levin lost with the collapse of the Soviet Union.

"Planet Pomegranate," Levin calls the connections made between lovers of the ruby fruit. I realized I'd joined this clan after my article on Levin appeared in *Orion Magazine* in Nov/Dec 2005, and people from all over the country and abroad began writing to me and sending their pomegranate art. What happened next can only be explained by the fruit's seductive powers. Richard Ashton, of Oak Creek Farms in Texas, Googling pomegranates, read *Orion* online and decided that he, too, would find a way to help Dr. Levin publish his full-length punicological study, *Pomegranate*. The last time Levin had seen his own work was when he'd passed the hand-typed, onion-skin pages, to botanist Dr. Bill Feldman from the University of Arizona, whom he'd taken on field trips in Turkmenistan in the 1990s. Translation funds had dried up and Levin thought his chance for publication lost, but now, thanks to Ashton's dedication, Texas A & M Press will publish the book. The Texas pomegranate grower also put together a small, practical guide for growing pomegranates, *The Incredible Fruit*, forthcoming from Texas A & M.

One last note: I asked Gregory to send photographs. As you will see, in his youth the pomegranate hunter did not so much resemble a Chekhov character as a young Jack London setting out for the wild.

I would like to thank Margaret Hopstein, Tashkent friend, gracious and intelligent reader and translator, who transformed a difficult Russian text into *Pomegranate Roads*. Botanists/biologists/fine stylists Marilyn Cannon and Elmer Dudik helped all along the way with careful, critical reading. The Texan dedicated-to-pomegranates, Richard Ashton of Oak Creek Farms, gave much support. Thanks also to David Karp, 'fruit detective'; Jeff Moersfelder and Joe Wertheim of U.C. Davis; Bill Feldman of the Boyce-Thompson Arboretum; Robin Beeman, editor par excellence; Shir Shnaider Kamhi from the Israeli Embassy in Washington DC and Dr. Ephraim Lansky of Rimonest; Muhabbat Turdieva, always willing to respond; Dr. Makmud Isar in Turkmenistan; Michael Morey for everything, Marilyn Kinghorn for proofing, and Maureen Jennings for her invaluable assistance in final editing. Lastly, *Orion* editors Hal Clifford and John Galvin, who coaxed out meaning for my article on pomegranates and kept me on the quest that led to *Pomegranate Roads*.

Barbara L. Baer
Forestville, California, 2006

Turkmenistan in Central Asia

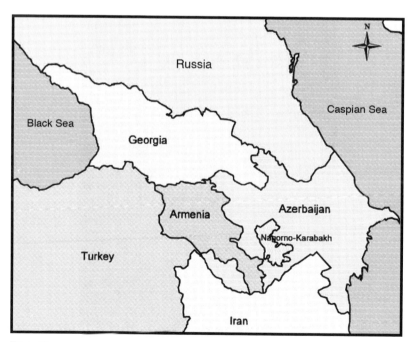

Trans-Caucasus

Levin in the Ukraine

# CHAPTER 1

# My Name Means Exile

**Permit me to introduce myself.** I am Gregory Moiseyevich Levin. I was born in Leningrad in January, 1933. My Jewish name is Girsha—its variants are Girsh, Gersh, Gershon, Gershom. In Hebrew "Gershon'" means "an exile." I have been given a somewhat symbolic name.

My grandfather named my father Moses after the prophet, the lawgiver to our people. I was named after the prophet's son. I named my own son Alexander (Iskender) after Alexander the Great who shook the world. In my family, we always took a serious approach to names.

Every author is a person with a mission. My life has always been connected to the Third Day of Creation when the plants came into existence, and studying them has been my mission. From childhood on, most of my days have passed searching for secrets under the Tree of Knowledge. In the vocational school before going to university, I worked part-time in a chemistry laboratory. My professor insisted that I choose chemistry, but I chose biology. My academic title has been Candidate of Biology, Senior Researcher. In Israel where I now live, a Russian Candidate of Science degree is the equivalent to a Ph.D. I am now retired.

The book you will read is a montage of my life and work with plants, especially with pomegranates, in the former Soviet Union.

Some chapters have mostly scientific content—may my reader forgive their dry language and know that later I will tell you stories about my expeditions and adventures. I sat down to write relying on my memory because I had to leave my reports and materials in Garrigala, Turkmenistan, my first Turkmen "dwelling." The brightest events remain in my memory, however.

The line from a nursery rhyme comes to mind. "I was born to be a gardener." Gardening is the most ancient profession. We know about tending the Garden of Eden. Life has tossed me about in the world, but I always remained myself, always with an inner direction independent of society around me, always fascinated by the world's garden. Age does not quench my thirst for knowledge. Whether this is good or bad, I do not know. It is the way it is. Writing intrigues me, though I wasn't born in search of self-expression but rather to be captivated by various biological theories and the plants around me.

I was eight years old when I planted my first garden without any instruction. The year was 1940. After the end of the terrible Soviet-Finnish war that took 300,000 Red Army soldiers' lives, the Soviet Union occupied the Karelian Isthmus, in formerly Finnish territory. As a child I saw the Mannerheim Line, a very strong reinforced concrete defense line with numerous pillboxes. My father had fought in the war, and after it was over, he worked as the director of a bookstore in the ruined Finnish town of Teriyoki—nowadays this town belongs to Russia and is called Zelenogorsk. The bookstore was in one of the two buildings that had survived. Behind the building was a big yard with a swing. There I decided, without being told by any adults, to plant potatoes next to the swing, and next to potatoes, I re-planted Hosta, an ornamental plant.

I have witnessed truly horrific events during the twentieth century, a century of bloodshed, misery, losses, deportations. Some call them "historic events." Whatever they've meant for historians, certain events are etched in my memory forever, starting with the unbearably cold winter of 1941–42 in Leningrad during the German siege.

My father's brother, Grisha Levin, a soldier and machine-gunner, was wounded and taken to a hospital operating in a school building next to the Marynsky Theatre. From the meager rations allotted him, he managed to save several bread crusts that he gave to my mother and me when we visited him. He was killed later in the suburbs of Leningrad, as was my father. As was Boris, his other brother. My own father's last "assigned residence" is a communal grave in Kolpino, near Leningrad. The poet Alexander Mezhirov writes: "Crowded near Kolpino, we were dying of friendly fire." Nicolai, a fourth Levin brother, a machine operator at Leningrad Metal Works, died of starvation at work, next to his equipment. Of the five brothers only one, Alexander, lived through the war. He was a foreman at a Kuibishev aircraft construction factory, and, as such, he got a waiver that kept him from the front.

All five brothers and two sisters were born in a little place in the Moghilev district of Byelorussia called Propoisk, which means, "Drink-All." When I was only two or three years old, my uncle Grisha took me to Byelorussia to see my grandfather. I remember a large garden wet from the autumn rain, the bare branches of apple trees. I remember a shed with a heap of Antonov apples, and my grandfather's water mill in a suburb of Moghilev. The Soviets renamed the village Slavgorod, "Glorytown."

The Great Patriotic War. The siege of Leningrad. One million Leningrad residents starved to death during the siege. I remember artillery shelling, the dead and wounded, the starving people on the streets that severe winter. In the beginning, school wasn't cancelled; they even gave us a plate of thin soup with about ten grains of millet floating in it. Soon that ended, both the soup and the schooling.

That winter was so cold that my ears, nose, hands were frost-bitten. To get water I had to walk several kilometers to the Neva River carrying a two-liter container to where a hole had been hacked through thick ice covering the river. There was always a line of people waiting to get to the hole.

I remember that we had a *bourgeuika*, an iron stove, for heating our room. There was no fuel. I had to chop our furniture to feed the fire. Hunger was the constant reality. Each person got 125 grams (4.4 oz.) of bread a day. Once on the way to the store for our rations, a big boy attacked me and stole the bread that I had just bought with our card.

My mother fell ill with scurvy. At the crossroads of Sredny Prospect and Third Line Street, peddlers were selling bunches of pine twigs. I cut the needles into a washbasin and added crystals of lemon acid. When the tincture was ready, I gave it to my mother. The concoction helped her to stay alive. My memory of that winter in besieged Leningrad is like a sequence of movie frames: the rhythmic metronome of the radio on the loudspeaker, broken into by occasional breaking news about approaching airplane bombardments or artillery shelling; the cold early dusk and the night hours; how water splashed from my container and at once turned into ice on the sidewalk. I remember corpses lying at the gates and right on the streets. Once I saw a partly covered corpse and recognized Vasia who lived in our house two floors below. Shortly before the siege he took my book to read, *Novellas from the City of Archangelsk*. I remember during artillery fire, a young girl was leaning tightly on her boyfriend or companion when the shells were flying over our heads. A woman's body on a stretcher was carried past us, and instead of a pelvis, the body had ragged bloody flesh. I remember at early dusk, a dying man on the ground trying to take food cards from another dying man.

In early spring, after the hard first winter in the besieged city, grass emerged in the Smolensk Cemetery. I collected nettle leaves and dug out dandelion roots. They made good soups full of vitamins that helped Mother recover from scurvy. At the Voencomat Military Committee, we were handed the notice informing us that my father had been killed in action. The destiny of Leningrad had been partially decided on the Pulkovo Heights. My father was killed in one of the decisive attacks there commanded by Georgy Zhukov, the future Marshal. (After the war ended, when I was already at college, I came to the Pulkovo Observatory located in the midst of a blissfully peaceful landscape.)

When a bomb exploded next to our house, our room on the sixth floor lost all its windows. Mother and I were evacuated to the Yaroslavl district. Getting there was not easy. As passengers died, their bodies were taken from the coach. On a ship crossing Ladoga Lake, we were bombed from the air. After we transferred to the train, we were again bombed from the air, but when the German planes dove at us, our experienced train engineer accelerated the train forward and backwards. From Yaroslavl, our train was ordered to go on to Stalingrad, but Mother and I managed to stay in the Yaroslavl district. Stalingrad soon after became the battleground of the century.

We received the last postcard from my father after he was killed. I was in second grade. Every day I harvested wild berries and mushrooms in the village where we were living near Yaroslavl. I explored the entire environment of the village, learned about the many plants, about when to pick ripe berries and mushrooms. It was there in that village, when I was 11 or 12 years old, that I developed my passionate desire to truly study nature. There were few books in the village. The Tree of Knowledge seemed to me the most desirable tree in the world.

I did not accept my father's death and continued to hope that he would come back alive. In 1945, the Leningrad Division was returning from Germany via Baltic territory to Leningrad. As I watched the soldiers walking past me, I felt certain that he would be with them. The miracle did not happen. Only after that, I stopped waiting, but I often heard myself repeating the lines from a Flemish book of the nineteenth century, *The Glorious Adventures of Tyl Ulenspiegl*, about a boy unjustly losing his father: "Claes' ashes were beating in my breast. His ashes beat against my heart." I vowed to transform evil to good. In 2002, honoring their memories, I placed the documents of the four dead Levin brothers in Yad Vasem Memorial Center in Jerusalem.

My mother and I returned to Leningrad after two years. We lived in the same street on Vasilievski Island but in a different house where we settled in a tiny room in somebody else's apartment. This nine-square-meter room was home for mother and me through all my schooling, vocational school, then college, always in that room. From there I made my trips to the suburbs, to the world of forests and swamps.

After the war, my mother worked two shifts, 16 hours a day in a cement factory in Leningrad. One day her exhausted girlfriend stumbled and fell down into liquid cement where she drowned at once. From the dismal reality of the time, I escaped into books that described the dedicated knights of science who challenged the dullness of ignorance and discovered new horizons. Louis Pasteur, I.N. Mechnikov, Luther Burbank, even I.V. Michurin were interesting to me. In *Uncle Vanya*, Anton Chekov wrote: "When there is no real life, one lives by mirages." I read more about scientists, especially biologists. I could read about, if not yet live, their adventures.

My biology teacher, Anatoly Michailovich Petrov, noticed my passion for plants. He chaired a circle of young naturalists in the Kirov Palace. We went on excursions and wrote reviews of our outings. He permitted me to check the schoolwork of younger students. "Little botanist" was my nickname at school. I learned that in 1918, during the Civil War, my teacher had fought in the Karakum Desert, in Turkmenistan. Later, when I was doing my post-graduate work in that same area, I mailed parcels of snakes and tortoises to Anatoly Michailovich Petrov.

In other ways, by the time I was 18, I was no longer a *tabula rasa*. The war that cost the country 28 million lives, among them two million Jews, were my growing-up years. I already had my own world inside myself. At 18, I may not have understood the many complexities of Soviet life, but I understood the main givens of our society that was totalitarian. That I was a Jewish boy who was born and would live uneasily in that society, and who would have to struggle to reach his goals. I knew my goals were to get an education to become a scientist.

I need to mention that early in the 1930s my grandfather was declared a "kulak." His property was taken and he was sent to labor camps to die there. Two of my mother's brothers, Nikolai and Robert, perished in labor camps as well. Robert was just beginning his career as a linguist, specializing in Arabic languages. Later, during "the Great Rehabilitation," as it was called, my mother was summoned to "the Organs" where she was informed that her brother Robert had been rehabilitated. They also told her the name of the woman who had reported on him.

I worked as a gardener in an industrial metal factory in Leningrad. I was sent to harvests in Kazakhstan's collective farms. In 1950, I found myself in the Kabardi Mountains on my first expedition as a plant-hunter. Our professor, Dr. Yury Ivanovich Kos, had us collect samples for the botanical garden of the newly opened Moscow University. We went to the Adir-su Gorge located on the border with Svanetia, the highest, most mountainous part of Georgia.

During that first expedition I remember the melody and a line of an old camping song: "Towns and countries, parallels, meridians are passing by." Particularly the last words, "meridians are passing by" stayed with me. Nowadays Kabardi is again Kabardi-Balkaria, as it had been before Stalin ordered Balkarians forcibly removed from their native land to Central Asia during World War II.

Not long afterward, I worked as an agronomist in a fruit collective in Daghestan. My interests evolved and changed. My first plant passion was *kniajhenika*, aka *polenika* or *mamura*—a dewberry, a small graceful plant with fragrant berries. I studied apple trees. I was most interested in the phenomenon of neoteny—premature blossoming and fruiting in polycarpic trees, tree-like perennials that bear fruit numerous times.

Osip Mandelstam called the twentieth century "a wolf-hound." His words have a deep sad meaning for me. I am thinking of the three revolutions, two world wars, numerous smaller wars, famines that came five times to ravish the Soviet Union. Famine had frequently occurred in Russia in the previous centuries, too. During the Civil War, during the famine in Odessa, my grandmother starved to death together with her two daughters, my mother's sisters. It was a miracle that the 1941–42 famine in besieged Leningrad spared my mother and myself. After that war, the times were not easy, either. People starved for a long time.

In the twentieth century, our Levin clan was reduced to me. All the men died before they had a chance to father children. My mother became a young widow with a terribly hard life alone, working to support me. I remained the only surviving male. Fortunately, my wife and I have a son Alexander and daughter Elena, and their children.

As Alexander Blok wrote: "Those who were born during calm years do not remember their way. We are the children of fearsome years in Russia, we are unable to forget anything."

When I was younger, life appeared to be so many adventures, odysseys, a chance to learn about various civilizations, to fill oneself up with diverse impressions. Being older, I'm able to reflect and rethink. Past events and people remain alive for me. I sometimes contemplate the question of one's inner potential, and how fully it gets realized. I probably should have searched more inside myself. Yet I'm inclined to think that one should not look back, or one may suffer like Lot's wife and Orpheus. In turning to see your city, turning to see your Eurydice, you may lose everything.

I think a romantic's life is more exciting than a pragmatist's life, even though I probably shouldn't make generalizations. Lately, I've been noticing another interesting phenomenon. In my past, especially my childhood years, time went so slowly! When I now look at the same duration, it appears to have flown. I know that I changed gradually, turning into a man on the road. Could it be the inheritance of my nomadic ancestors, a descendent of Abraham's wandering tribes?

Finally, for many long years, I studied the pomegranate in all its aspects, selecting the best varieties, and then distributing them to many countries. Every new region was a test site, a proving ground for my observations, a kind of laboratory. I attempted once to calculate my mileage through all those years. I had to give up trying; there were too many miles in my life. In Russia, I explored Leningrad,

Moscow, Yaroslavl and Tambov regions, Karelia, the city of Omsk. In Belarus, I explored the Moghilev region. In Ukraine, Kiev and Uzhgorod, the Khmelnitsk region and the Crimea. In the Caucasus, Kabardino-Balkaria, Daghestan, Azerbaijan, Armenia, Georgia, Abkhazia, Adjaria. In Central Asia, Kazakhstan, Turkmenistan, Uzbekistan, Karakalpakistan, Tajikistan, Kirghizstan. I made many return visits to those areas. I was also in Istanbul, Turkey.

Since the early 1960s, my home base was the VIR Turkmen Experimental Agricultural Station, called Garrigala. For many of those years I was chairman of the laboratory of subtropical fruit cultures at Garrigala. Of course I was not alone in this effort. Half of the work lay on the shoulders of my wife, Emma Konstantinovna. Like all women, Emma had a more practical approach.

Our homemade small universe in Garrigala, created by my wife and myself, collapsed a decade after the end of the Soviet Union. When we left, the germplasm, or living pomegranate collection at Garrigala, had 1,117 samples, or accessions. It contained more varieties than any other in Central Asia and the world. What I cared most about were the excellent varieties of pomegranates and other subtropical fruits we collected and developed there. I am gratified that the efforts of my mental and manual work now exist in many places around the globe.

There was a lot of sadness in leaving Garrigala and coming to this new land of Israel where, thanks to the meandering path of my destiny, I am learning the art of living in one of the world's noisiest crossroads. Israel is a very beautiful country. Generally speaking, I never saw a country that was not beautiful. Here the transition to colder temperatures is smooth. Flowers are blooming in the cities and along the roads all year round, making a deep impression.

I have undoubtedly developed a somewhat romantic attitude towards Israel, and I hope that my new universe will emerge here in my historical land. But where else could my wife and I have gone? No home remained for us in Russia any more. Russia did not need me. Leningrad—St Petersburg—had become distant. In any case, I lost Russian citizenship after the collapse of the Soviet Union when I automatically had to become a citizen of Turkmenistan. Russia lost a significant amount of its intellectual resources when it broke apart.

Not long ago, I phoned the Garrigala station to wish a happy birthday to Natasha Burnasheva, the last remaining employee of the 1960s generation to which I belonged. She had retired but not left as Emma and I had. Natasha informed me that the new director had ordered massive bulldozing around the entire territory of the station. All of it was gone, the bushes, trees, old houses, including the one my family lived in for 40 years. The order from the Turkmenistan Ministry of Agriculture was to plant vegetables instead.

All my life, I've had a slightly mystical feeling about woods and forests. This feeling remains strong in Israel where the majority of woods are not wild but planted. Still, each time I'm in a forest it's a celebration for me. I live in Haiarkonim—Street of Greengrocers. There is some symbolism in that sequence of names and places, a continuum with being nomadic in time and space that began long ago. I now have a test site in the Bet Shemesh (House of the Sun) woods where I observe plants and have published three articles on the subject of succulents.

In my life, I've been in the temples of various religions. I've watched religious services of Christians and Muslims. In Israel for the first time in my life I saw religiosity that emphasized outward expression. I think that believing should be an intimate, interior process. It should not need any advertising or intermediaries, middlemen communicating with the Superior Being. Actually, I consider everybody is a believer, even those who think that they are not. I am a secular person. I don't attend religious services. Science is my temple. To be a scientist is to be on a mission. Nothing is better than that. Understanding the world is one of our deepest needs.

Once in Haifa, we were shown the Bahai Temple. They didn't require intermediaries with the Superior Being. They are like Sufis who deny ritualism. I am attracted to Buddhism for its nonaggressiveness. In my roaming over Muslim Central Asia, in contacts with sincere believers, I saw that usually they were highly moral people.

In my opinion, the role of the Jewish people in global civilization was determined by its own heretics. Their thoughts and writings have been preserved through the centuries by a people driven from everywhere, scattered everywhere. Among heretics I place Abraham, Moses, Jesus, Einstein, Bohr, Freud. The list goes on and on. Heretics leave their teachings, their paradigms, to us.

When I was a child, I wanted to be 'like everybody else,' which meant avoiding the painful humiliations that Jews suffered. I might have wished this, but my life did not work that way. I remained a man of the Diaspora, a traveler in time and space, a voluntary Ahasuerus, (the Wandering Jew, not the Persian king also named Xerxes,) a curious tourist, a Russian Jew, a Russian speaking Euro-Turkmen who for 40 years lived in Turkmenistan like a particle of sand carried by desert winds. I have been a participant of the twentieth and twenty-first centuries, involuntarily brought into history, like everybody else, like my near and dear, like those whom I knew and those I did not, like all my colleagues, and others who left no trace in history, but without whom, perhaps, there could be no history at all. We are not just dust in the wind. It depends on everyone.

It is natural that I am pondering now what will remain of me after my timer goes off. It is never late enough to die, but after 70 years, it is hard to die prematurely, as A. E. Bovin, a talented journalist and the Russian ambassador to Israel, has said.

One thing I didn't learn was how to make money. I gave my entire life to acquiring new knowledge without learning how to convert knowledge into cash. I chose the wrong profession for that. I think I have the right to say that my quest was to explore the harmony of the world through science. In the beginning, I perceived that science was the best way to escape from social problems. Later I saw that science was joy itself. Of course, from time to time, I had the vanity and desire to push science ahead myself. But always more interesting was to understand phenomena. I felt a scientific detective's excitement in decoding the unknown. My entire life was in a garden, the way I wanted it, and under the canopy of Science, attentive to nature and her riddles. I remember a woman, a journalist, writing that understanding the laws of nature can be considered a specific intellectual sport—that science was a cross-word puzzle. A.A. Abrikosov, Nobel Laureate in physics, said that a scientist is never bored.

Leningrad poet Alexander Kushner wrote, "One does not choose one's times, one just lives in them and dies." Kushner's lines are never absent from my thoughts, a refrain that comes and goes. You will hear me repeat it and read it between the lines.

Contemporary life is a grand artifact. In some lives, there is a death of conscience. But there are cases of the immortality of conscience. Academician Nikolai I. Vavilov was such a conscience, an immortal, fearless conscience. His life and story run through my own and inspires it.

I am now on the shore of the Phoenician Sea re-reading M. Bulgakov's *Master and Margarita*. Though I don't have itineraries ahead, the habit of racing against time has remained as strong in me as ever. I think of M. Lermontov's line: "And he, rebellious, is seeking for a storm, as if it is in a storm that peace could be found."

I could never accept the role of a lame duck. I need to be in harness and find a reason for living. Thus, when this book was suggested, I began to write about myself and the plant so important to my life. Of course you know it is the pomegranate.

*Pomegranate Roads* is my first experience in personal writing. I understand that writing is not my specialty, yet I still hope that you readers will not be left indifferent, and that you will read to the last page.

The words of the singer Andrey Makarevich are my cue to begin. "We are fortune hunters looking for a bird of ultramarine color," he sang. I, too, saw a bird of ultramarine in the West Pamir Mountains, and, elsewhere, I saw a blue pomegranate.

# CHAPTER 2

# First Pomegranates; under Stalin's Shadow

**Stanislav Ezhi Letz wrote that one can close one's eyes to reality,** but not to memories. Memories are coordinates of the past that return us to the days and pages of life gone by. Being born in January, I live under the sign of Aquarius. My gem is the garnet; in Russian, *granat* is also the name for a pomegranate.

The pomegranate is "the subject of my little story," as Anton Chekhov said about a certain work. I first saw a pomegranate fruit long ago, in 1938, in my native city of Leningrad—now St. Petersburg. After marching in a celebration rally on Revolution Day, November 1938, my father and I stopped at our relatives' home. A tightly-diapered newborn baby, my second cousin Esther, lay on a large bed, and next to her I saw dark-red and dazzling pomegranates. I thought they tasted amazingly good. Later, I learned they were brought from Azerbaijan. My vision of them is as fresh as if they were in front of me now. Three years later my father was killed defending the approach to Leningrad.

My second exposure didn't happen until 1957, after I had graduated from Leningrad Agricultural College, Fruit and Vegetables Department, called today St. Petersburg Agricultural University. Academician N.I. Vavilov taught at the college. I will write more of this man later. I chose apprentice work as an agronomist for a large fruit-growing *sovkhoz* in southern Daghestan, a mountainous country in the North Caucasus.

During the hard years of the agricultural collectivization, many kulaks were deported to Death Valley—that's what the Sharakhoon Valley was called because of rampant malaria in southern Daghestan. There, in the valley of the Gurghen River, the exiles built a *sovkhoz* named after Gereihanov, a local communist leader who had been killed by enemies of the regime. The *sovkhoz* gardens spread for six miles and occupied 1,976 acres. At the end of the 1940s, on Stalin's order, many subtropical cultures were planted in the northern lands, outside their natural habitat. Nothing good resulted from this order. In the Gereihanov *sovkhoz*, all that remained were several pathetic pomegranate bushes, annually frostbitten. During my three years there none of them even bloomed. They were miserable parodies of pomegranates.

After working for three years in Daghestan, the Department of Fruit Cultures of the VIR in Leningrad twice rejected and finally admitted me to post-graduate study. Out of 50 applicants for the post-grad opening, I was the only one who had passed all the entrance exams with good grades, but they still denied me full-time status. I was rejected because of Jewish quotas, not because I wasn't qualified. Only through friends and the help of Professor N. G. Zhuchkov was I finally admitted. There I was assigned a research topic connected with the biology of apple trees in Turkmenistan. Professor Zhuchkov told me that years before he had studied the so-called Baabarab apple trees in the Central Asian republic. He advised me to give them my special attention.

In March of 1961, without further hesitation, I left for the Kara-Kala settlement, which is now called Garrigala, home of the Turkmen Experimental Agricultural Station, where I would spend the next forty years. At that time, there were 14 VIR stations— Experimental Research Institutions for Land Management—in different parts of the USSR They might be compared to the USDA facilities in the United States. It will be easiest if I refer to them simply as agricultural stations.

That little settlement was located in the valley of the Sumbar

River in the southwestern Kopet Dag Mountains, located in the middle of nowhere—only desert and mountains around. The hot breath of the Karakum Desert reached it from the north. The Iranian Kopet Dag portion of the Turkmen-Khorasan Mountains was to the south. The Caspian Sea lay some 200 miles to the west.

The mountain settlement had come into being through the efforts of Academician Nikolai Ivanovich Vavilov, who was the founder and the original director of the chain of agricultural institutions. Nikolai Vavilov, who was born in 1887 and died after being tormented and tortured in prison in 1943, was a world-class scientist who generated and passed on numerous ideas. He formulated the theory of centers of origin of cultured plants. Another of his contributions was the theory of homologous series in genetic variability. His disciples became scientists in their own right while investigating and realizing his ideas in the USSR and elsewhere worldwide. Descending generations of scientists continue to honor him by following in his path.

Vavilov was highly appreciative of the southwestern Kopet Dag region where he'd first visited in 1916 during his expedition to Persia, present-day Iran. In 1936, he wrote, "This region is one of the global centers of the formation of our most valuable dry subtropical varieties. Great numbers of varieties and sub-varieties of almonds, walnuts, pomegranates and grapes are concentrated here as natural growth in the Sumbar River Valley and along its gorges and canyons."

Vavilov himself chose the Garrigala location for the station. The area on the eastern frontier of the Turkmen subtropical zone had a median annual air temperature of 60° F—equal to the median annual air temperature of the earth.

When I came to Asia, to the isolated and far-away land of Turkmenistan, a new world opened for me. During my first years in Central Asia, everything seemed exotic. I did my best to learn as much as I could. Later, the region became my world.

The Sumbar Valley region was a real laboratory of nature, rich in abundant wild relatives of cultivated plants, and a subtropical climate. Ancient history was all around, so the many archeologists who have worked here often told us.

Vavilov valued in people what he called "the genes of decency." He was a scientist who was highly perceptive about human qualities. Under his tutelage, only top scientists were invited to work for the agricultural institutes. Like Vavilov himself, many of these scientists were later crushed in the Soviet terror. Figuratively speaking, he had to stand like a gladiator against the malicious, shameless, unscrupulous dishonesty of Lysenko and Lysenkoists in their personalized hate campaign. Looking back on history, one shouldn't ever underestimate the importance of a single man, for good or for evil. If Vavilov had remained as the head of the agricultural institutes, everything would have been different. Blocking his way was Lysenko.

Timofey D. Lysenko, a peasant's son with little education, was one of the dark presences of the Soviet epoch. In my opinion, his peasant mentality influenced his views. Certainly his lack of scientific training contributed to his doing such harm. Glorified as the "peasant genius" and "barefoot scientist," he rose to power in the 1930s as a Stalin favorite and served as part of the propaganda campaign for collectivization.

Lysenko was a proponent of the eighteenth century theories of Lamarck, who believed that organisms could inherit acquired traits. (Lamarckism was a purpose-driven theory, as Darwin's natural selection was not.) Lysenkoism rejected genetics. It basically denied all the achievements of modern evolutionary and genetic theory known in Stalin's time. Lysenko himself represented all the undereducated people who believed that environment, not heredity, determined biological characteristics. What turned this chapter in our history of science truly sinister and tragic wasn't the ignorance of one man and his simplistic, anti-evolutionary doctrines, but that he gained Stalin's support.

Early on, Lysenko's unscrupulous adherents craved power. "Gray" people—talentless, undereducated, mediocre people—who were placed in power by Stalin surrounded him. They had a slogan—"This Babel should be destroyed!"—a kind of code to identify scientific positions such as Mendel's and the genetic role of evolution that were more advanced than their own, as something from the outside, something to trick common people. Stalin imposed Lysenko's theories on Soviet scientists who either had to renounce genetics and everything modern or suffer the consequences. This was ideology over experimentation. At that time when reason itself collapsed, all the human sins of envy, hatred, ignorance and lust for power came to the fore, all of them lined up in the campaign against western science and its scientists who were stigmatized as "Mendelianists/Morganists."

It's almost impossible to estimate the damage Lysenko caused. He deprived people of the opportunity to pursue their work. Biological sciences that had advanced with the rest of the world were set back decades. Because of Lysenko, thousands of excellent, talented biologists, botanists and agronomists were sent to prison or labor camps and destroyed or killed. Nikolai I. Vavilov himself was tragically one of the terror's victims. His death is a story that still is painful to recall.

Vavilov fought to preserve Soviet biological science against the Lysenko-dominated camp, but Stalin's sinister power was always just behind. It was truly a battle between universal human arche-types: a teacher of justice and truth against a horde of liars; innocent victims against executioners; light against darkness.

In 1940, Academician Nikolai I. Vavilov was arrested on false information. He died in a prison in Saratov in 1943 after 400 inter-rogations. What interrogators did to that elderly and sick man before he died is not anything you want to think about.

In 1994, co-workers and disciples published a book of essays by 78 scientists dedicated to Vavilov. These researchers were and continue to be the glory of Soviet science. Eighteen of them had been arrested during Stalin's terror. Nine perished in prisons or labor camps. One, a woman, was shot. Besides these, many starved to death in Leningrad during the siege. But even as they starved in that freezing winter, they saved Vavilov's collections of seeds and cultured plants.

Even today we do not know the names of all the scientists who perished in the gulags or prison torture chambers. I believe it is inevitable with time that the names of most Vavilovists will become known and rescued from obscurity. They took the high road to scientific truth and did not leave it under all of Lysenko's and Stalin's threats and tortures. I was privileged to have met 27 of these heroic scientists.

Stalin is no more. But like tyrants before him, his ominous figure casts a long shadow over our century, our country, like the period of McCarthyism in the United States. Maybe James Joyce was right when he called history a dream of a drunken pub owner. Michel Foucault considered twentieth century history as a form of insanity. Maybe they were right, especially when applied to the times of Lenin, Stalin and Hitler. In the endless fight between good and evil, why has evil so often won?

A scientist does not live in a cocoon. His epoch intrudes into his life, into the very objects of science, into the world of plants, their studies and application thereof. Now, for myself, even distanced from that era, it still looks horrifying and harmful, as well as inhumane, illogical and senseless. But then I am reminded of the words I wrote earlier, "One does not choose one's times. One just lives in them and dies."

I knew someone in the agricultural institutes who wrote accusations regarding Vavilov. G. N. Shlikov was not completely talentless; he just did not have morality, and he took sides with the prevailing Lysenko adherents. He did what he was told, fulfilled their ignoble tasks. He lived long enough to suffer himself. The Soviet terror caught him in its net. He was exiled to the camps. He survived and returned, got a Doctor of Science degree. It was only after the younger generation of the post-graduates started making noise about his past that he was finally barred from any of our agricultural institutes.

The whirlwind times of the 1930s came as far south as Garrigala in Turkmenistan. There, Mikhail Grigorievich Popov, a well-known Russian botanist, became the first director of the Garrigala experimental station. He's been called "a courageous knight of science." Popov tolerated no inertia in botany; he was never satisfied with the status quo in science. He formulated ideas that were always revolutionary. In his youth, Popov initiated the concept of the Ancient Mediterranean area. He became an advocate of the theory of the hybrid emergence of varieties. For many years he worked in Central Asia, discovered and described numerous varieties of flora.

M. G. Popov's destiny was also to be caught up in the Lysenko outrage. He was arrested in 1933. Some time later he was released from prison, but was forbidden to work in Leningrad and Moscow. He was sent to work in many god-forsaken places far from where he had done his most important work.

I was still a post-grad when I arrived in Turkmenistan. In the beginning and throughout my 40 years there, I traveled a great deal in the southwestern Kopet Dag learning to know its gorges. It was there that I came across my first wild pomegranate thickets. At that time I did not know how fateful that discovery would be.

Nikolai Mikhailovich Minakov was one of my companions at that time. As a young man, he had traveled widely in the Central Asian mountains assisting Vavilov. He had fallen in love with Olga Mizghireva, a young lab technician. Later she would direct our station. But no matter how far one might be from Moscow, Stalin's arm reached even farther. Minakov was arrested in 1938. For ten years he was a prisoner in northern labor camps. His daughter Natasha was born after he was arrested, and only met her father for the first time when he returned from the camps.

I was fortunate that the cataclysm of Soviet history had calmed down by the time I began my work. As a new generation of 1960s scientists came to positions of leadership, we had the fortune of working under conditions of greater independence that also gave us inner freedom. We were able to make contacts with visiting elite scientists, intellectuals and artists. My wife and I opened our home to many visitors from distant cities, something that would have been impossible in the earlier times under Stalin.

In 1964, when I began my work on pomegranates, the collection of varieties at the Garrigala station consisted of 64 accessions or samples. By the time I left, we had 1,117. Until the Soviet Union collapsed, I participated every year in expeditions to discover and collect pomegranates and other subtropical fruit in various regions of Central Asia, in Turkmenistan, Uzbekistan, Tajikistan, Kirghizstan, Kazakhstan and the Trans-Caucasus, in Georgia, Azerbaijan and Armenia. I worked on other fruit and plants as well. I remember collecting wild sugar cane plants in the Amu-Daria River meadows in Turkmenistan to send to Cuba for crossing with the Cuban cultivated sugar cane.

As I planned expeditions throughout my life, I was inspired by Nikolai Vavilov's letters that he'd sent to friends and colleagues. His energy and daring, his understanding of struggle and success, kept me on task. Life is short, problems are innumerable, but the horizons of the unknown are wider yet. The world should be studied by roaming across it.

It so happened that I witnessed the prime time and the sunset of this little center of culture at the very edge of Soviet influence. Garrigala was a tiny oasis in the spiritual wasteland of our Soviet times. It was our wonderful new world. Being young, we did not miss everyday comforts.

Of course, living at the station had its limitations. I missed playhouses and theatre, art galleries, newly published books. But we had radio, and new movies were played in our little town. Many of us had good books, tapes and LP records. By 1980 or so, TV came to Garrigala.

In my post-graduate years, not only my work preoccupied me. As a young man, I frequently enjoyed a pair of attractive eyes meeting mine. There were many acquaintances, many a romantic rendezvous. As the poet Yuri Levitansky wrote: "Each man chooses for himself his own woman, religion and road." My search for the right woman took a long time and ended with marriage. My choice was a good one. My wife Emma Konstantinovna and I have been married for 43 years. We met in Turkmenistan, "meeting on a distant meridian." She was an intern from the Ashgabat Agricultural Institute. She entered my life to stay and became my devoted friend and assistant. We have two children. We had much in common in our families' pasts. Like mine, her grandfather was "exposed as a kulak" during the Soviet collectivization of agriculture. Her childhood trauma had been the Ashgabat earthquake in 1948 that had caused the deaths of a third of the city's population and killed some in her family. Mine had been the siege of Leningrad. We both were passionate about working with plants.

My young wife Emma assisted me in numerous cataloging inventories. Before the pomegranate became my center of interest, there were apple trees. What amazed me most were the various and abundant anomalies and abnormalities occurring in apple trees in our region. These were probably the results of radio-nucleotides that fell as rain and contaminated the soil after the largest Soviet H-bomb was exploded above ground on the New Land Archipelago in the Arctic Ocean in 1951.

After I completed my post-graduate research, I was offered the position of directing the laboratory of subtropical fruit cultures at Garrigala. Emma had by then graduated and joined me. She worked there for 40 years, too. In the laboratory, she studied grapes. She added 200 varieties to the collection from our Central Asian branch near Tashkent, Uzbekistan. Later she worked in my laboratory, adding to its collections and researching the germplasm of persimmon, fig and olive.

There were many new plants I learned about in Turkmenistan. It took several years to familiarize myself with the collection's varieties of pomegranate, fig, olive, persimmon and jujube. I created the jujube collection at Garrigala.

As time went by, visiting students and post-graduates brought us up to date with the outside world. They entertained us with new samizdat-recorded songs. Self-published bards were becoming famous among intellectuals. We were all passionate about photography, slide shows, etc. We had local poets. Some of us were good at painting. Many years later, former employees of the station and former post-graduates published memoirs describing their research and expeditions.

In the good old Soviet times, expeditions were sent here from all over the Soviet Union. Members of the Academy of Science visited Garrigala. Academicians, professors, all manner of scientists, bird-watchers and butterfly-watchers, writers, poets, journalists, artists, painters, tourists and many others came here. Some of those contacts turned into friendships that lasted decades. I'm writing this story in part because the history of our scientific colony mustn't be lost to the sands of time.

In Russian and then Soviet times, we have had periods of rich collection and preservation of species and times of loss. Perhaps a summary description will help my readers understand the periods of historical change that I've described above. As best I can, let me recall the dates from memory and give a name to these eras much as we Russians named periods of our literature.

**The Beginning,** 1894–1921, from E. Regel to N. I. Vavilov, who came as a young professor from the town of Saratov.

**The Golden Age,** 1921–1940, when Academician Vavilov became director of the VIR, the Soviet Agricultural stations across the vast land.

**The Regression,** 1940–1951, Institutes taken over by Lysenko's protégées.

**Awakening,** 1951–1961. Post World War II, Academician P. M. Zhukovsky became head of VIR.

**The Silver Age,** 1960–1980, Academician D. D. Brezhnev headed VIR (do not mistake him with Leonid Brezhnev who ran the Soviet Union at the time of the so-called "stagnation.") Numerous expeditions explored the regions of the USSR and other countries, with a great increase in the number of species in collections.

**New Regression and Degradation,** 1990s, as the Soviet Union began to fail.

**Collapse,** at the end of the twentieth century and the beginning of twenty-first century, with sharp reduction in financial support, loss of five experimental stations located on the territories of the

New Independent States, and shrinking of scientific work.

When Turkmenistan became an independent country in 1991, new governmental agencies began supervising us, replacing the USSR They changed the name of the station. The new government of Turkmenistan apparently did not need science, and in particular, punicology—the study of pomegranates. They gradually stopped financing the station. We faced the worst financial crises imaginable.

I was shaken by the situation in science after the Soviet Union collapsed, caught unprepared for everything to fall apart. In the wake of the collapse, the Soviet Union and the agricultural institutes abandoned their scientists and researchers at the experimental stations. We were assigned to the emergent sovereign states, left without any protection, without any possibility of continuing to work the way we'd been trained and needed to work. We no longer had a center. At that time, I wrote letters to the directors of the agricultural institutions but everyone was so preoccupied with their own immediate problems that I received no response.

Some Western sponsors attempted to help our station. Professor Victor Fet from the University in West Virginia came to visit. Years before, Fet had begun his career in Garrigala in the Sunt-Khasardag Zapovednik where he was an arachnologist, a specialist in spiders. Some money was also raised by Professor Bill Feldman, director of Boyce-Thompson Arboretum in Arizona. The little money we received couldn't ameliorate the desperate situation. The crisis became systemic, and nothing could save our station.

The entire network of stations was absolutely penniless. Gone were the expeditions. Publications ceased. Many of our scientists and researchers passed away—let us hope to a better world—or were fired/forced into retirement. Others emigrated, as I did. It was unbearably painful to witness the collections perishing, dying. I had given 40 years of life to create and study them. (I was so healthy there that for 30 years I was a blood donor.) On the threshold of my fourth quarter of a century, my face and body carry the traces of my past life, the inevitable abrasions and lacerations. Wherever I am in this big and stormy world, Garrigala will be with me forever.

My bluebird of good fortune happened to be a ruby colored fruit. The pomegranate became my sacramental culture. I've never been sorry that I chose the pomegranate. As for the bluebird, I saw it years later in the southwestern Pamir Mountains in Tajikistan.

# CHAPTER 3

# A Brief Description of the Pomegranate

**Some scientists consider the pomegranate a tree**, but it is a bush. Trees are characterized by having only one trunk. No single-trunk pomegranate has ever been found in a natural environment. In nature, the pomegranate is a bush. Domestication is another matter. Here we are free to do what we wish to shape a pomegranate as a small tree. Pomegranates can grow 13–20 feet high, occasionally to 23–33 feet. In very severe natural environments, on rocky slopes for example, you'll find low, even creeping bush varieties. Cultivated dwarf varieties can be as small as ⅔ foot or as tall as five feet.

These widely different-sized bushes are all in the same pomegranate sub-family. In the past, we believed the family had only one genus—*Punica granatum*—within which there were two kinds: the common pomegranate and the Socotra pomegranate that grows only on Socotra Island off the coast of Yemen. My research concluded that these are two separate genera. The Socotra is now called *Socotria*. It is a genus that has only one species. The common ancestors of the two genera come presumably from the original continent of Gondwanaland, a single landmass that floated apart and eventually became Eurasia, Africa and part of America. The pomegranate family is a kind of an evolutionary dead-end. We find other families like that, with two genera in them. 5.5% of all fruits that we know about belong to families with limited possibilities.

Let's think for a moment of the relationship between plants and humans. If we date back modern man for about 800 generations, we know our ancestors were first hunters and gatherers and later became animal husbandry men and agriculturalists. Over thousands of years, people learned how to use a variety of plants on all continents, at most latitudes. Through experience, agriculturalists selected plants for their usefulness. There are no less than 250,000 varieties of flowering plants or angiosperms. Of these, at least 75,000 have edible parts for humans. Through all our history, however, only about 7,000 plants are known to have been used as food. An even smaller one percent of the earth's flora has been domesticated. Of those, fruit-bearing plants were among the most important because fruit provided easily digestible carbohydrates, vegetable oils and other properties. Fruit became especially useful as humans evolved a smaller jaw—more brain size, less jaw. We needed a new food that didn't require the kind of chewing our ancient ancestors were capable of. Fruit fit our needs perfectly.

Fruits are not so much energy foods as they are a source of vitamins and other biologically active substances. There is a tremendous diversity among wild and cultivated plants in various regions and climates. Fruits grow on trees, bushes, shrubs, and semi-shrubs, as tree lianas and as herbaceous lianas. There are herbaceous polycarpic plants that fruit a number of times in their lives, and monocarpics that are fruitful only once.

Tropical and subtropical zones produce almost 60% of all global fruits. Their variety is also greatest in these zones, where pomegranates definitely have their place. We don't know when the first pomegranate appeared, when we might celebrate a first birthday, but we do know that at least 50–70 million years separate that date from the present day. Fossils of myrtles, in whose family the pomegranate belongs, were found on the ancient continent of Gondwanaland. We can't know for sure that the first pre-pinicoids (the ancestors of pomegranate) didn't appear with the evolution of this family. There were probably pomegranates on the shores and islands of the shallow Tethys Ocean. After Gondwanaland broke apart, the plants migrated across Indostan along the belt of *alpian orogenesis* (the area in the processes of mountain making) to occupy an area that shrank greatly with the Ice Age.

Presently, in its wild form, the pomegranate is distributed in Eurasia from the Balkans to the Himalayas. It is most frequently found in lower mountainous regions. I've estimated that about 1,000,000 individual bushes are growing in natural conditions, with about 100,000 on the territory of the former Soviet Union.

Domestication of the pomegranate came about with the Neolithic revolution. Most probably, domestication happened in several places and at different times about 5,000 years ago in the countries of the fertile crescent in the eastern Mediterranean and over 2,500 years ago in Armenia, the Crimea, etc.

The pomegranate entered into cultivation first with people who collected the fruit, then practiced a basic horticulture by selecting the best samples. Domestication followed. In some areas of Central Asia, I've observed this process going on now, as wild pomegranates are introduced into cultivated areas.

The pomegranate preserved some of its tropical features and acquired a number of evolutionary new characteristics that permitted it, after being domesticated, to cross over the borders of its natural area into new locations.

Optimal conditions for the pomegranate exist in Mediterranean type, arid climates with high exposure to sunlight, annual total precipitation of 7–22 inches; mild winters with minimal temperatures not lower than 10°F. Just as optimal conditions vary geographically, so do soils, microclimates, and the level of agro-technology, and each of these factors play a role. Early sweet varieties don't need as much heat. They can be grown in locations farther north than sour-sweet varieties of pomegranate.

Under domestication, the pomegranate grows in all subtropical and tropical countries in the wide area of 41° northern latitude to 41° south—a pan-tropical and pan-subtropical area of cultivation in moderately warm zones. In Western Europe, decorative pomegranates have been grown at 51–52° north. Amateurs in the Soviet Union grow pomegranates in areas 44–50° north but pomegranates are not frost-resistant. In Central Asia and in China, gardeners cover pomegranates to protect them from the cold. Dwarf varieties freeze at 19°F, soft-seeded at 10–12°F, hard-seeded at 0–3°F.

Pomegranates are heat-loving plants that can bear high air temperatures of 114 or 118°F combined with hot dry winds, though flower petals burn under these conditions.

In natural environments, pomegranate plants can live to 100–150 years and more. In cultivated gardens, there are known plants 300 years old.

Domestication covers 114 provinces of 33 regions of all six bio-regions of earth. The pomegranate easily runs wild and becomes naturalized under the right conditions.

Presently pomegranate centers are in Iran, Afghanistan, Turkey, India, Pakistan, Israel, the Magreb countries of North Africa— Algeria, Tunisia, Morocco; also Spain and the U.S.A., particularly in California.

Pomegranates tolerate excessive calcium and thrive on alkaline soils, though they grow well in a variety of soils. The pomegranate is essentially a *glycophyte* and can stand some salinity, but they don't bear fruit well. My expeditions took me to pomegranate gardens in Kizil-Atrek, southwestern Turkmenistan, an area of secondary salination resulting from salty underground water layers rising to the surface. The pomegranate gardens there had been badly affected and were doomed to die. The best soils are deep, richly aerated loam and sandy gray loam.

Pomegranates have a *xylopode*, a peculiar underground woody stem that is their storage organ to supply nutrients, in particular to restore the aboveground bush if it dies back after a cold winter, and as the plant ages. The *xylopode* is formed during the first years of the pomegranate's life.

Pomegranates bear fruit in their second or third year. Full fruit bearing comes at 7–10 years. (In the natural environment, the process is slower.) In optimally favorable conditions, one plant can produce 440–660 pounds of fruit annually. For many years I was monitoring wild pomegranates in the Narli Gorge of the Chandir River in the southwestern Kopet Dag. There were several bushes that bore over 1,000 fruits each.

The pomegranate fruit is a kind of berry with a hard pericarp—a seed vessel—enclosing the fruit's contents, of a crimson or yellow-green surface. The pericarp's inner side is covered with albedo, spongy matter, most likely needed to protect the inner fruit from heat and, to a lesser degree, from cold. In different varieties, the color of the skin varies from white-greenish to dark purple, almost black, but most frequently on a scale of reds. The fruit contains many, many seeds or arils with a sour-sweet, sweet, or, rarely, an acutely sour taste. The color of the pulp varies from transparent, almost white, to dark crimson to nearly black.

Occasionally you can see *metaxenia*, when in the same pomegranate, among the seeds of the same color (for example, crimson) there are one or several seeds of another color. This is believed to have happened because of double fertilization from another variety.

Early sweet varieties ripen in August-September, but the taste quality doesn't last—it's gone by December. The later varieties ripen by October-November and preserve their excellent sour-sweet taste until March-May. They can be preserved in storage with regulated temperature and humidity. The size and mass of fruits vary significantly from just several ounces among decorative dwarfs to 1.1 pounds and up to 2.2 pounds. In extremely favorable conditions, you might have a fruit weighing 4.5 pounds or more.

Fruit size depends on the variety and the time the trees blossom. The largest fruits develop during the first wave of blooming. Each successive wave of blossoming results in smaller and still smaller fruits. This phenomenon is especially predictable in the northern areas of wild or domesticated pomegranates. Pomegranate bushes blossom several times a year in warmer areas nearer the tropics. This has to do with photoperiodism in the pomegranate, especially present in regions closest to the tropics. In the regions along the northern border of its natural area, pomegranate blossoming is reduced to one wave.

In the southwestern Kopet Dag, when the pomegranate is domesticated, small fruits make up one-third of the entire number of fruits on the same bush. In natural populations, small fruits are rarely observed. They do not reach ripeness and thus the rind does not break, and the fruits remain on the plant till the next year where they can be used as seed reserves for preserving the variety. Small fruits do appear at the end of blossoming when the weather is very hot, and the pollen-carrying insects have disappeared. The small fruits are the result of self-pollination. I verified in numerous tests that seeds of such small fruits that stayed a year on the plant grew well, and that they carried all the features of the mother plant.

Some pomegranate varieties are vulnerable to sunlight intensity. Sunburn lowers their marketable qualities because it stops growth and development of the fruit. For our collections at the Garrigala station, I selected a group of varieties resistant to sunburn.

A cracking rind in pomegranates is a biological device allowing seeds to distribute themselves in natural environments. In cultivation, cracking isn't a desirable feature; gardeners try to find a way to prevent cracking and the huge harvest loss that results. In the wild, it is quite rare to come across populations of pomegranate whose skins do not crack. After a number of years of research and studies, we were able to select for varieties that did not crack.

When cultivated in a domestic site, pomegranates are a water-loving plant. With sufficient irrigation, an acre of pomegranates can yield 4–10 tons of fruit. I met Tajik gardeners who insisted that pomegranates need as much water as cotton and cucumbers.

In nature, however, pomegranates predominantly grow in conditions of water shortage. For this reason, scientists consider the pomegranate to be a xeromesophyte, an intermediate between xerophytes adapted to life and growth in dry environments with water shortage, and mesophytes, plants that grow in habitats with medium rainfall.

Nowadays, water is considered a precious resource; many countries have shortages. We see this happening already in the Middle East where economizing water and/or water recycling has become an imperative. There are forecasts that in the future, battles will be fought over water. In Israel, where pomegranates are an important crop, drip irrigation and recycling has reduced water use by 30%.

My predecessors who studied pomegranates in Garrigala concluded that pomegranates had great potential to grow in dry regions and non-irrigated forests or orchards. I also suggested that we try irrigating pomegranates with sewage water. We selected certain pomegranate varieties and transplanted the saplings to an area where we could observe the results. The experiment was successful. To the best of my knowledge, we were the first to try irrigating pomegranates with sewage water.

# CHAPTER 4

# Punicology, the Science of the Pomegranate

**Botany is the scientific study of plants,** with subdivisions denoting special fields such as Rhodology (roses and rosehips) Batology (brambles like blackberries and raspberries) Agrostology (cereals) and Pomology (apples and pears.) Punicology studies the pomegranate, which comes from the Latin *Punica*, meaning pomegranate. Punicology includes morphology, physiology, anatomy, ecology, geography, varieties, agricultural technologies and other scientific and applied disciplines relating to pomegranates.

Academician Nikolai Vavilov led the way in Soviet punicology. In 1916, during his first trip to what today is Turkmenistan, he familiarized himself with pomegranates in the wild and those that were domesticated in the Sumbar River Valley in the southwestern Kopet Dag Mountains. Vavilov strongly believed that the pomegranate should be studied for many reasons and that we must preserve its biological diversity in collections and in the wild. While he was still alive, he oversaw the creation of a pomegranate collection in Azerbaijan and in 1930 in Garrigala, in Turkmenistan.

In Vavilov's expeditions across Central Asia, the Trans-Caucasus, Iran, Afghanistan, China, Syria, Spain and many other countries, he gave his attention to pomegranate distribution and how it was domesticated in each region. In a letter to A. D. Strebkova, an early pomegranate researcher at the Mardakyan Agricultural

Station in Azerbaijan, Vavilov wrote that he had a "soft spot" for pomegranates, and it was time to begin writing his philosophy of pomegranate evolution. In another letter to Strebkova, he wrote that Turkey's Minister of Farming had visited him and seen on his desk the blue pomegranate fruit from a seedling in Strebkova's selection. The Minister was amazed to see a blue pomegranate.

Three generations of punicologists worked in the Soviet Union during the twentieth century developing Vavilov's ideas, creating collections and profoundly expanding our knowledge of the pomegranate. I belong to the third generation. From 1930–70, punicologists A. D. Strebkova, A. G. Pruss, B. C. Rozanov, M. G. Popov, N. K. Arendt, N. I. Zaktregher, E. A. Gabrielyan-Beketovskaya, and O. P. Kulkov, worked in various research institutions of Central Asian and Trans-Caucasian Soviet Republics.

Among these outstanding botanists, let me just mention Mikhail Grigorievich Popov, who I referred to earlier as a "courageous knight of science." Popov, who became the first director of the Garrigala experimental station in the 1930s, believed that the pomegranate family was almost monotypical and was endemic to the area of the ancient Mediterranean. In the pomegranate genus, he wrote, one was distributed in that ancient Mediterranean area, the other on Socotra Island. The richest, most perfect varieties were concentrated in Iran, Asia Minor and the Trans-Caucasus. In these major areas, pomegranates appear to have developed on the basis of their wild biological diversity.

In our region of Central Asia, Popov found that many of the abundant pomegranates in the Kopet Dag Mountains resembled the ones in Iran, but that in neighboring Tajikistan, wild pomegranate diversity was not as great. Domesticated varieties there were numerous but not so good as in the wild. In the years since Mikhail Grigorievich Popov did his work, pomegranate research continued at Garrigala and elsewhere. Soon I will tell you about that rare and distant Socotra Island and the discoveries I made concerning their unique pomegranate.

# CHAPTER 5

# Centers of Pomegranate Origin and Variability

**Nickolai Vavilov was the first to formulate** the theory of the centers of origin and the genetic diversity of cultured plants. He theorized that plants had established (definable) borders extending from the centers where they first appeared, and that they had specific traits and relationships to other plants found only within those centers. He identified the primary megacenter of cultured plants to be in the Near and Middle East.

Pomegranate genetic resources were not equally distributed around the world. I followed Vavilov's general identification as it applied to pomegranates, theorizing that the primary endemic pomegranate megacenters were also in the Near and Middle East. I then identified secondary megacenters in the Mediterranean and Eastern Asia, and tertiary megacenters in America and in South Africa. I also identified a number of microcenters within the three top megacenters.*

From these central areas, pomegranates were introduced into many adjacent areas by means of seeds, grafts, cuttings and saplings, creating the basis for forming local varieties. Over time, in many

---

* Inside the megacenters, I identified 13 microcenters in Malaysia, Iran-Afghanistan, the Trans-Caucasus, Central Asia, India-Pakistan, the Near East, North Africa, the Pyrenées, the Balkans, China, Japan, North America, Latin America.

secondary megacenters, valuable varieties developed. Significantly later, the process was repeated in the tertiary megacenters. The fact that secondary and tertiary centers were far away from the primary megacenter is not unique to pomegranates. Practical work with sub-tropical and tropical plants has shown a similar history with coffee, cinchona, cocoa, sugar cane, banana, walnut, cashew, agave and many other plants that developed far from their centers of origin.

North Africa once had the most fertile land in the Mediterranean area. Carthage was famous for its excellent pomegranates. But during the Roman era, intensification of agriculture, cultivating virgin lands and cutting down woodlands resulted in soil erosion. Vandals, and then nomads, completed the destruction of the flourishing agricultural center. Thus Carthage, famed for pomegranate culture, perished and lost its prominence.

In the first 30 years of the twentieth century, interest in pome-granate cultivation dwindled, both in the Near East and North Africa. Citrus growing took its place. Developments in pest and disease control, however, have brought pomegranates back into favor. The India-Pakistan region has been increasing areas to grow the ruby fruit.

Among endemic centers, I should name the Kandahar oasis in Afghanistan, which has the largest pomegranate fruits in the world. The Tagob Valley, also in Afghanistan, is famed for seedless pome-granate varieties. Soft-seeded varieties are grown in many regions of the Middle East and the Mediterranean. Kazake-anor, the most frost-resistant pomegranate variety, originated in Uzbekistan.

In Turkmenistan, there are gorges characterized by concentra-tions of pomegranates with some very definite features or qualities. In the Narli Gorge of the Chandir River Valley in the southwestern Kopet Dag, you'll find the largest concentration of wild, small-fruit pomegranates with tiny, lightweight seeds. In a number of other gorges, along with small-fruit pomegranates, you'll also find non-cracking varieties, and varieties with perforated seeds. I explored all these wonderful gorges and will write in more detail about my expeditions and discoveries there.

# CHAPTER 6

# Decoding the Socotra Pomegranate

**There is no royal road in science,** only continual work, collecting material and analysis, and then, perhaps unique, individual and frequently intuitive solutions emerge from the work process. This is the story of decoding of the Socotra pomegranate.

It had been long known that the pomegranate genus had two branches: the common pomegranate and the Socotra pomegranate, found on the island of that name off the northeastern coast of Africa. It was thought that this more ancient, primitive variety, was the original, and the common pomegranate was its derivative.

Socotra Island was quite remarkable. Characterized by an ancient, tropical flora with numerous endemic plants existing only there and nowhere else, the island had varieties preserved in isolation for eons.

The common pomegranate, as we've seen, had a wide distribution. The Socotra pomegranate was discovered in 1880 by J. B. Balfour, an English botanist. Much later, A. Radcliff-Smith, who worked in the Kew Royal Botanic Gardens in the U. K., wrote to me that on the vast territory of Socotra Island, he was able to find only four old Socotra pomegranate trees. Nor had their varieties fruited in Kew Botanic Gardens, where they had been propagated by cuttings and grafts.

After much research, I came to understand that the Socotra

pomegranate was a separate, independent genus. Paleo-botanist Inessa Alexeyevna Shilkina, in an article published for the Russian Academy of Sciences, St. Petersburg, wrote that she found fiber tracheids (water-conducting cells) in Socotra pomegranate wood. Fiber tracheids are not found in any other plants of the Myrtle family to which the pomegranate belongs. This is a remarkable fact: the Myrtle consists of 16 genera, 427 species, 10,257 varieties. Among them, only one, namely the Socotra pomegranate, has those fiber tracheids!

I needed to prove my idea that the Socotra pomegranate and the common pomegranate were not brothers but, rather, cousins at least twice removed. Not a single scientist would trust the evidence just by my saying it was so, but Professor Vasily Vasilievich Nikitin, Chair of the Agricultural Institutes Herbarium and Flora Department, offered his help and assistance. By the way, my wife Emma had been his student before we married, and he had wanted her to remain working for the department he chaired.

For my research, Professor Nikitin requested the Socotra from the largest herbarium, the Kew Royal Botanic Gardens. He invited me to familiarize myself with this rarest variety. After that, at my request, a live Socotra pomegranate from Kew Gardens was sent to me at our Turkmenistan experimental station. Unfortunately, when the parcel with the Socotra pomegranate arrived, all the leaves had fallen. The plant did not survive the shipping. A second order was placed—the same thing. It was not possible to get a live Socotra plant for our station. There was a silver lining. We had the material for morphological and anatomical research.

A young researcher, Lena Sokolova, studied the anatomical peculiarities of the Socotra pomegranate. Lena was a student of another woman of scientific prominence, Professor Veronica Kazimirovna Vasilevskaya. Many years earlier, when I had been a post-graduate at the Turkmen station, I had shown Vasilevskaya many amazing examples of morphological anomalies that I had observed in apple trees. Significantly later, we arrived at the understanding that the abundant deformities resulted from the tests and explosions in the early 1990s of the largest Soviet H-bomb on the New Land Island in the Arctic Ocean. Rain had carried radioactive nucleotides over the enormous territory of the U.S.S.R, causing these side effects. I never saw published materials on the influence of the tests on humans, but undoubtedly people were subjected to effects of heightened radioactivity in the same areas as the apples had been.

In studying the Socotra pomegranate, I enjoyed the assistance of a number of scientists from various research institutes of the Soviet Union. I studied the unique morphology, in particular the leaves, blossoms and fruit. Other fine scientists studied chromosomes, seeds, pollen composition. The results of the collective work showed that, in all its features, the Socotra pomegranate was a narrow *paleoendom*, where a genera or species stands apart from the other endemic species of a region. The Socotra was a cul-de-sac in the evolution of the pomegranate family, with significant differences at all levels from our familiar, common pomegranate. We were convinced it was a monotypic genus with only one species, and it was given the name *Socotria*. Thus, the common pomegranate became orphaned and had to be considered as a monotypic genus as well.

*The Botanical Journal* published the results of the research that I co-authored with Lena Sokolova. I published two subsequent articles describing *Socotria* as a new genus, and giving the evolutionary particularities of the pomegranate family. A quarter of a century has passed, and only a few botanists have recognized *Socotria* as a new genus. Never underestimate the power of inertia!

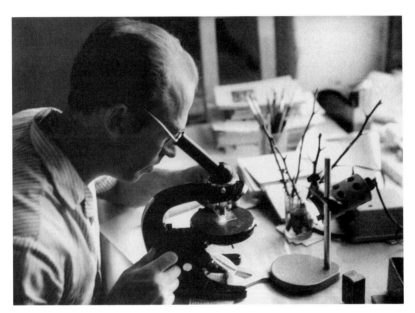

Levin at microscope

# CHAPTER 7

# The Pink-petaled Pomegranate

**When I'd just begun working at Garrigala in 1961,** and was reading records of work done in previous decades, I came across Ludmilla V. Klimochkina's report from 1937. She described her expedition to the Chandir River in the Narli Gorge where she found a wild pomegranate with pale pink blossoms.

The Chandir River Valley, 31 miles south of our Sumbar River Valley, had a much warmer climate than where we were at our station. During relatively severe winters, the coldest temperatures were not so bitter. Of the many gorges in the valley, five were called Narli "pomegranate" Gorge. The Submar River Valley also had five gorges called Narli. Apparently these early visitors didn't stretch their imaginations when it came to names. There were more wild pomegranate plants in the Chandir River Valley, as would be expected from the warmer climate, than in the Sumbar River system.

When a delegation of botanists came to visit Garrigala, one of them asked if I knew of L.V. Klimochkina. I answered that I was familiar with the name but that I hadn't seen her publications. The botanists said that Mila—short for Ludmilla—was a very beautiful woman who had studied pomegranates at our station. Still I let it go until years later when I went exploring the gorges of the Chandir River Valley and reached the very

Narli Gorge where she had found the pale pink-petaled pomegranate bush.

That gorge had abundant pomegranate growth, over 1,000
bushes, certainly the most extensive local population in the southwestern Kopet Dag. In the 40 years since Klimochkina had seen
it, the unusual pink-petaled population had grown significantly
larger; where there had been one plant there were now several
attached bushes. That pink-petaled mutant form did not fruit. Still,
even the fruitless form was valuable for its rare characteristics, and
so I added it to the collection.

Much later, at the end of the twentieth century, an international
delegation visited the station. The specialists came from Uzbekistan,
Italy, Tunisia and the USA. We visited the Narli Gorge. After I told
my story to Professor Mars, a renowned pomegranate expert from
Tunisia, he became excited about the pink-petaled pomegranate
and took several cuttings to his country. Unfortunately, there's no
resolution to this story. Dr. Mars and I corresponded and mailed
publications to each other until I moved to Israel. He then stopped
writing. I concluded that his reason must have been my relocation
to Israel and his situation in Tunisia. I always feel sadness when I
think that prejudices turn out to be stronger than scientific interest.
I also wonder about his pink-petaled pomegranate.

Southwest Kopet Dag

Southwestern Kopet Dag region

# CHAPTER 8

# Mountainous Areas and Seismism

**Many of the areas where pomegranate distribution** goes far back in time had high seismic activity. Earthquakes were common. Turkmenistan in the Central Asian ranges fit this profile. The 1948 Ashgabat earthquake killed 80,000 people, toppled buildings, tore houses and trees apart and left meter-wide splits in the earth.

The first time I experienced the ground shaking was as a child in non-seismic Leningrad, the winter of the siege. A big bomb exploded next to our house, shaking it so hard that my mother's bed rolled on its little wheels. She was lying in it, sick with scurvy.

The first real earthquake I experienced was in Samarkand, Uzbekistan. My hotel began shaking, the electric bulb hanging from the ceiling oscillating, water in the pitcher on the table shaking. People ran from my third floor downstairs. That earthquake was about 3.5–4.0 on the Richter scale. Seismologists in Garrigala told me that there were an average of 200 quakes annually in the region and since that first one, I have experienced so many that they became like second nature for me.

I'm not digressing here from pomegranates. Earthquakes frequently result in plants being covered by rocks of various sizes. When this happens, pomegranate branches manage to root themselves with time, and a "derivative" develops, a bush connected to the mother bush whose branches, trunks or saplings are pressed down or covered with dirt that eventually forms additional roots or whole root systems.

In mountainous regions there's frequently high natural radiation, an important cause of intensive mutation of plant varieties—the explanation for the various spontaneous mutations in wild pomegranates.

# CHAPTER 9

# Mutability, Varieties and Selection

**All organisms, all kinds of plants and animals** are mutable to some degree. My research has shown mutability in all aspects of the pomegranate, from size and appearance of the plant, to its shoots, leaves, buds, flowers, fruit and seed pulp, to complex functions within each of the organs. In the 1930s, in the Azerbaijan agricultural department, A. D. Strebkova worked with seedlings that were grown from the wild pomegranate seeds collected in the gorges. She had pomegranates with weeping forms, with blue pulp and pericarp, a yellow-petaled variety and much more.

The southwestern Kopet Dag happened to be a real refuge for recessive pomegranate mutations that we added to our collection at Garrigala. There were dwarf mutants, creeping plants that never stood up, dense-branched plants I found only once on a steep slope in the mid-Sumbar River area. We found those with narrow leaves, multi-colored petals, with varying pericarp, seed vessel and seed mutations, and with rinds that didn't crack. The explanation probably is that the area was part of the center of origin, a center of biological diversity. Additionally, like many of these centers, it is characterized by naturally high background radiation.

My experiments inducing mutation with 10 kilo Roentgen radiation doses resulted in some peculiar results: a creeping seedling with shoots growing in different directions and not growing upwards.

It was fruitless, unfortunately. Another resulted in a transgressive wild type—the expression of an inherited gene in combination with other genes in a hybrid offspring—with harder, heavier seeds than normally found in nature.

Among the seedlings that grew from seeds soaked in a super chemical mutagen, I selected one whose qualities included early ripeness, in the beginning of August. The fruit was a beautiful festive pink with sweet pink pulp and soft seeds.

Beyond the many mutations, we can observe over 50 centuries of purposeful selection in various pomegranate 'birthplaces,' resulting in hundreds of local varieties.

In 1930, Nikolai Vavilov spoke to the Fifth International Botanists' Congress in Cambridge, England. The next year he published his ideas in an article, "Linnean Species as a System." In this important writing, Vavilov defined a plant variety as a natural complex, genetically diverse, that included categories of various sizes. A variety was an offshoot on the evolutionary tree with mutable morphology and physiology that in the process of its formation was connected with a certain environment and area.

It is apparent from his intricate definition that plant varieties and sub-varieties can vary significantly in their size, their composition, their place among others and their origin. I arrived at this conclusion after studying pomegranates in many different locations where they were growing wild, in agricultural centers and in herbarias.

Looking at a pomegranate population in the wild, you will find a large group of plants that are similar in some of their features, but have a variable morphology. The structure of the variety is not revealed with clear morphology. Variability within that variety—its subspecies, or intra-species variety, if you like—does not reach a sufficient degree of organization in the pomegranate to be considered taxonomically a different variety. There may be morphological differences in flowers or fruit or leaves or shoots but to make the claim of a subspecies, an entire area must show these differences. In nature, different varieties can be found in mixed populations occupying the same area.

Nikolai Vavilov said that selection was actually evolution determined by man's will. The beginnings of selection go back to the tribal collectors. Our nomadic ancestors were quite observant. Those who settled in the regions of wild pomegranates noticed the good, large fruit that tasted delicious and had soft seeds and other desirable properties. They selected these ripe wild fruits to plant in their settlements.

As time passed, our ancestors understood that seeds retained the pomegranate's qualities. Over even greater time, observance and curiosity led to developing methods of pomegranate propagation, primarily by taking cuttings from seedlings, or young branches from established trees, and planting them in loose soil that were kept moist. This was an enormous step in the development of horticulture and gardening. It simultaneously permitted people to grow orchards with the best fruit trees in their area, as well as distributing those best varieties to distant places.

The first attempts to classify pomegranate varieties were made in ancient times, as new varieties appeared—the product of culture, monuments to the difficult work of orchardists of the East. Pomegranate varieties were so attractive and delicious that conquerors were amazed at the achievements of people they considered primitive. These conquerors carried the best pomegranate plants to their own countries as trophies among their booty. As a result, many pomegranate varieties of today were distributed all over the Middle and Near East after military campaigns.

Trade caravans along the Silk Road also participated in the distribution of new pomegranate varieties. From Southern Arabia, pomegranates were brought to Egypt. In the ninth century B. C., pomegranates arrived in Carthage. The fruit brought from Carthage came to Italy in the second century B. C. Phoenicians brought the fruit to Spain, as did the Moors in later times. (By the thirteenth century, Ibn-al-Avam described ten varieties grown in southern Spain.) China received its ruby treasures from Samarkand in the second century B. C. In the era after the Conquistadors, pomegranates appeared in America.

Excellent pomegranate varieties came about mostly through amateur selection as local 'specialists' continued to create even better new varieties with qualities they wanted. This has always been true—there seems to be no end to the achievements of gardeners and scientists. Charles Darwin wrote that frequently a flower or fruit would be declared perfect, only later to have it surpassed. Nikolai Vavilov's phrase expresses my meaning: "Selecting varieties is the result of human intervention in the nature of plants."

The 1930s became a new era of pomegranate scientific selection. At our station, it was primarily the result of the great women scientists. A. D. Strebkova handed the seeds of the first generation of soft-seeded pomegranate hybrids over to Olga Fominichna Mizghireva, our director. Mizghireva grew seedlings that she planted on a special test site. This was slow work that required great patience. In the end, her patience was rewarded.

How do features change in the pomegranate selection process? A most important pomegranate feature for the Soviet territories was resistance to frost. Our patient researchers increased the number of pomegranate hybrid seedlings with good frost resistance in the second generation and more in the following generation. Like many other fruit plants, old pomegranate varieties, usually the ones selected by amateurs, were better at passing their features to the next generations than varieties more recently developed through selection, a phenomenon that still remains somewhat mysterious.

When I was gathering, preserving and studying collections at Garrigala, it was not my goal to develop new pomegranate varieties. Working on selection is different and results take a long time. You need a particular personality with enormous patience and immense specialized knowledge to do that work. Nonetheless, there came a point where I had to study and resolve a number of questions on pomegranate biology that were directly related to selection. Briefly, here are some of my results.

As I studied the biology of the pomegranate flower, I found that it was characterized by protogyny, that is, the ability of its pistil, or rather its stigma, to mature and become susceptible to pollination earlier than its own anther, the pollen carrier, matured. This characteristic makes pomegranate flowers susceptible to pollination with the pollen from other flowers, cross-pollination, earlier than self-pollination. From this interesting discovery, we concluded that to hybridize we didn't have to alter the plant's own flowers, a labor-consuming task that involved removing the anthers. It was possible to cross-pollinate by just removing the closed sepals, the outermost circle of leaves that hindered access to the reproductive organs.

In the company of other punicologists, I often learned what I needed to do to solve a problem. I also appreciated the correspondence with amateur gardeners from around the Soviet Union. I was happy to send them seedlings from varieties we had developed, in particular the soft-seeded and relatively frost-resistant plants. Many of those amateur gardeners informed me of their results with a particular variety, so information went both ways.

Gardeners, who I hope are among my readers, are less interested in abstract concepts than how we grew unusual and special varieties such as the soft-seeded pomegranate, the black pomegranate, small fruit pomegranates and non-cracking types.

# CHAPTER 10

# The Soft-seeded Pomegranate

**When humans introduced pomegranates into culture** and began domestication, there was always a trend in selection to grow soft-seeded, less-seeded or seedless varieties. Soft-seededness is a valuable recessive mutation in pomegranates that has been known since the fruit was first grown. Theophrastus, the father of botany, wrote about soft-seeded pomegranates in Kylikya in Asia Minor. Gardens in ancient Carthage were famous for their soft-seeded pomegranates. Merchant ships delivered bushels of soft-seeded pomegranates to Rome. In modern times, famous soft-seeded pomegranates were grown in the Royal Gardens in the Tagob valley in Afghanistan and in Tunisia.

Selection of the recessive soft-seededness appears to be quite dependent on the environment where the pomegranates are grown. When, in the nineteenth century, soft-seeded pomegranates were planted in the Tsar's property in Murgab, present-day Turkmenistan, their distinctive feature disappeared.

The ancient gardeners were known to be great empiricists. If only we could observe them at work, we could learn a great deal about how they created the seedless varieties of pomegranate. We can speculate that ancient gardeners were drawn to the seedless mutation, and through selection using generations of seedlings, this valuable property was fixed and strengthened.

In some varieties, soft-seededness varies significantly from one year to another, probably having to do with whether they are cross- or self-pollinated. Some people have named this qualitiy *semi-softseeded* to signify that this feature varies annually.

During an exhaustive examination of wild pomegranates in the Sumbar River Valley near a deserted village, I saw several plants that had light, small seeds. One hundred seeds weighed about one gram. The seeds had apparent perforations, openings in their hard cover—features typical of soft-seeded pomegranate varieties as well, though their seed covers are not as hard as these were.

Professor B. S. Rozanov thought that soft-seeded pomegranates that appeared spontaneously and couldn't retain their soft-seeded feature in the natural environment happened because birds and other animals ate all the seeds. In my opinion, however, that explanation is a comfortable but not a reliable hypothesis—another useful expression provided by Nikolai Vavilov.

Our Soviet scientists worked for decades to develop soft-seeded pomegranates. Their fruits were large and beautiful, had pleasant sweet-sour or sweet taste, soft seeds and high juice content. There was one drawback common to all: they could not survive low temperatures. This feature limited their areas of distribution. In Central Asia, where pomegranates have always been sheltered by their growers, the vulnerability to cold of soft-seeded varieties did not matter as much.

During my years at the station, I selected and gave away the best soft-seeded pomegranate varieties from the collection to 50 scientific institutions in 16 different countries on four continents, including the U.S.A. The number of donated samples was significant. In essence, they constituted mini-collections of pomegranate varieties that included the best we had. Our Turkmenistan station gave many varieties to the USDA National Clonal Germplasm Repository, University of California, Davis, California. Our varieties make up 51% of the entire collection in Davis.

In 2005, Barbara L. Baer, residing in California and a great fan of the pomegranate, tasted the soft-seeded varieties sent by the Turkmen experimental station to the University of California at Davis. Jeff Moersfelder, greenhouse manager of the pomegranate collection in Davis, California, contacted me by e-mail to express his enthusiasm for our pomegranates. I later learned that after these tastings, anyone who wished could order cuttings. I was heartened to learn that our Turkmen varieties received the most requests which meant that in some years, they would be growing and enjoyed halfway across the world.

# CHAPTER 11

# The Black Pomegranate

**Years ago I read an article** in *Soviet Subtropics* magazine that in the 1930s, in Azerbaijan, a scientist found a pomegranate with black fruit. Naturally, I was very surprised. A black pomegranate, no less! I mentioned it at one of my meetings with Alexandra Strebkova. She told me there was such a plant in the Mardyakan collection in Azerbaijan, and that it was also being grown in Agdam, a regional station nearby.

In 1980, I visited the agricultural station and Agdam, a small town whose main product was a port wine of the same name, popular in the USSR. In a neighboring village, in a garden belonging to Azad Fakhradov, I found the legendary black pomegranate growing. It turned out not to be truly black but more a dark purple or dark violet. The fruits were not large, about one-half pound each, with pink arils. They were very sour, not tasty at all, but Farkhadov told me that *Kara-Nar*, the black pomegranate, was very good for treating diarrhea and vitiligo, a pigmentation disorder that leaves the skin with white patches.

Before I left Agdam, curiosity led me to visit a mosque for the first time in my life. People said the photograph on the wall was of the Prophet Mohammed.(!)

In Garrigala, I planted *Kara-Nar* seeds. The variety yielded excellent results as far as reproduction of its principal traits and turned out to show these traits when crossed with other varieties—that is, the black pomegranate traits were dominant. It was also resistant to frost during our southwestern Kopet Dag winters. In our station's collection, the fruitful *Kara-Nar* bush impressed visitors. I remember one Italian on an international expedition who circled the bush with his camera taking numerous pictures.

My most recent encounter with *Kara-Nar* came in Israel. In Kfar-Saba (Grandfather Village), quite close to where I live now, I found a hedge of pomegranate bushes. Growing on it were *Kara-Nar*, black pomegranates!

From my experience, I have concluded that *Kara-Nar* could be used as an excellent decorative variety, planted as a solitary bush, or in a group, as a border or hedge. And recalling Fakhradov's account, I hope the black pomegranate's medicinal potential will be explored, especially in Israel where the nutraceutical properties of pomegranates are being studied and put to use.

When I look back on my visit to Agdam, I wonder if that village survived the collapse of the Soviet Union and the wars between the Azerbaijan and Armenian armies fighting in Nagorno Karabakh. I wonder also if that mosque that I entered is still standing, and whether Azad Fakhradov and his family are alive. What I do know is that the black pomegranate made its way to other countries via our collection. On the bushes, the abundant glossy black fruit seen from afar make an unforgettable impression.

# CHAPTER 12

# Small-fruited and Non-cracking Pomegranates

**Inspecting local pomegranate populations in several gorges** in the southwestern Kopet Dag, I came across bushes with abundant small dark-red fruit. Those small decorative balls made a lovely picture against the background of the large rocks. Until the end of the year, their festive little balls decorated the plants like bright Chinese lanterns.

The small decorative pomegranate varieties could be used in city parks and yards because they are quite resistant to dust, gas, wind and various chemical waste.

In natural conditions, the majority of wild pomegranate varieties and many cultivated pomegranates have one unfortunate feature from our perspective: their rinds crack and split. It happens to a larger or lesser degree depending on the environmental conditions and on the specific pomegranate varieties. Cracking and splitting are an important biological mechanism for dispersing seeds, which I saw especially in late ripening and sour-sweet varieties. Some studies of pomegranate varieties have shown no correlation between the degree of pomegranate cracking and their ripening date, their taste, or the thickness of their pericarp. In any case, cracking is a negative feature for growing the fruit and bringing it to market or storing it.

Let me digress for a moment to pursue a more general principle concerning pomegranate seed dispersal. Fruits may begin to crack quite early, long before they are ripe. In the wild, especially in the years of drought, you might see cracked fruit as early as July to August. The same happens with pomegranates in gardens when they're not watered enough. When a pomegranate fruit cracks, its seeds (arils) fall to the earth. That's when dissemination begins; seeds are scattered, eaten by birds on the ground or on the tree. I've seen all kinds of animals eating pomegranate seeds, from mice to porcupines, boars, even cows grazing in the gorges in southwestern Kopet Dag. These animals carry away the pomegranate seeds, sometimes for dozens of kilometers. Ants and termites carry them away at ground level. Rain washes them down hills and gorges.

If seeds lie in favorable conditions, they may germinate the next spring. But once seedlings grow, these young plants survive only under extremely favorable conditions—conditions that don't happen often in arid zones. In fact, optimal conditions occur only three to five times every thousand years. Sufficient water is necessary for the first several years until root systems develop and can reach sufficient underground water for the entire growing season. When conditions are not favorable, seedlings die during their first hot, dry summer. My observations in the Narli Gorge proved to be a test site. I monitored plants there, and annually described their conditions, counting the crop on the test bushes, and observing self-seeded seedlings.

There are other reasons why the pomegranate seedlings did not grow every year. When there were a great deal of rodents, the so called 'mice years,' rodents devoured pomegranate seeds completely, and the next year there were no seedlings.

The southwestern Kopet Dag mountains frequently had severe winters that froze both wild and cultivated pomegranate plants. After such severe winters, the plants did not produce fruit for several years or had insignificant crops. During droughts, we observed a similar failure of plants to seed themselves.

We've digressed from the cracking pomegranate and now will

return to it.

When the skin is cracked, a fruit isn't as marketable. To prevent this from happening, pomegranates can be harvested before they ripen, which isn't the best choice. However, there are ways to ameliorate cracking without early harvesting. At Garrigala, we developed several pomegranate varieties that had minimal rind cracking.

In some varieties, the rind breaks only a little, and nowadays agro-techniques can regulate soil and air humidity, maintaining both, if possible, at the same level to ameliorate cracking. You want to have as little fluctuation in soil and air humidity as possible and a correct schedule of irrigation in the pomegranate orchards.

You can also minimize the cracking process through selection. You begin with varieties that resist cracking under variable conditions, though it's not often that these exist in nature. The Turkmen popularly classified desirable non-cracking types as *Donghuz-Nar* or "Pig Pomegranate." These trees had late ripening fruit, with quite thick red-green striped skins, fruit with very large red arils, an excellent sour-sweet taste. The fruit stayed good a long time on the tree. Frequently a Turkman went to the mountains to take cuttings from these pomegranate varieties to plant them in his orchard. Occasionally he dug out whole bushes to re-plant in orchards.

Photo: Anitra Redlefsen

# CHAPTER 13

## Propagating Pomegranates by Seed

**From the literature,** I knew that in a number of regions in India, pomegranates were being propagated by seed. I'd heard about this also in Turkmenistan, in the Konghur *kishlak* or settlement near the city of Meri—formerly Merv, the ancient city now entirely in ruins that Alexander the Great had conquered. Meri had a famed pomegranate culture.

When we approached Konghur, it was already getting dark. I was with my usual assistant and driver, Eghenmurad, a quiet modest man, who contributed greatly to journeys in Turkmenistan. He was a practicing Muslim, who began his day with a prayer. He then took the wheel and covered enormous distances. He arranged our shelter and nourishment so I wasn't distracted from work. Eghenmurad died several years ago. His heart gave out. He was four years younger than I was.

When we arrived in Konghur, we entered a house where a very old Turkmen was lying on a felt floor mat, too thin to be called a rug. His middle-aged sons were respectfully serving him while their old father told us about his quite poor childhood, long before the 1917 Revolution. Once, with his brother, he'd approached a *doovahl*, an adobe fence. Behind it, rich people were having a feast. They saw large and beautiful pomegranates. The diners ate the beautiful pomegranates and threw their seeds over the fence where the boys picked them up and planted them. The brothers became known for this pomegranate garden in the Konghur *kishlak*

Pomegranates that I saw still growing in the orchards of Konghur were the sweet, early ripening variety called *Ak-Donah*, very popular in Central Asia and widely distributed in neighboring Uzbekistan.

During my expeditions, I encountered seed propagation in other regions in Turkmenistan, in Tajikistan, and in Kazakhstan. Pomegranate culture in the traditional Central Asian way was determined by what is called a "continental climate"—severe cold in winter and very hot summers. Before winter arrived, people tied pomegranate bushes to the ground and covered them with a layer of dirt to protect them from freezing. In spring, they removed the dirt layer. To me, that was a lot of labor, but pomegranates are the favorite fruit among Muslims and can be sold at a high price, so hard work paid off.

When I came to Konghur again a year later, the old man was no more, but every household still had a large pomegranate orchard. The *kishlak* grew enough pomegranates to satisfy the appetite of the city of Meri. The gardeners stored their fruit in unheated structures on the straw-covered floors or on the shelves. The longer the storage, the more out-of-season they became, the more pomegranates cost at the market. Some pomegranates stayed fresh until May. It was curious to me that in neighboring villages and *kishlaks*, people were growing no pomegranates, only in Konghur.

### Unexpected Continuation

My reading on the subject of seed selection, combined with observations during expeditions across the Soviet Union, convinced me that there could be significant practical applications to propagating by seed. For ten years I studied the progeny of various self-pollinating varieties. Before the flowers opened, little gauze sacks—isolators—were put on the buds to prevent insects from getting into the flowers and bringing pollen from other varieties, ensuring self-pollination. In the autumn, the fruits were harvested.

Frequently, the self-pollinated fruits were smaller in size than the cross-pollinated ones. We collected their seeds and grew seedlings at a special test site away from our orchards. When this first entirely inbred generation of seedlings yielded fruit, all their seeds produced plants with genetic traits identical to the parents. This continued into the second inbred generation. The phenomenon appeared quite mysterious to me. It was an illustration of isogenesis, or identical genetic make-up that results from inbreeding.

I asked for consultation and assistance from embryologists at the Institute of Botany in Ashgabat, headed by the renowned N. S. Belyaeva. For six years these embryologists studied materials that I collected at the test site. Pomegranate embryology was still in its early stages. The experts told me that apomixis, exactly what we were doing by propagating plants without fertilization, hadn't been observed in pomegranates. I consulted with agricultural geneticists whose answers were also vague and imprecise.

Though it appears that further studies are required, I have arrived at several practical conclusions. With self-pollination, all features are inherited from the parent plant. From that, it followed that pomegranate varieties could be propagated by seed—as indeed, must have been the ancient practice of gardeners around the world. Secondly, it was possible—and easier, simpler, cheaper and safer—to preserve pomegranate seed collections rather than live plant collections of pomegranate varieties. But I needed to test this idea. I shipped 20 samples of our seed sets—the first and second inbred generations—for storage/preservation at the cryo-bank in Pushchino on the Oka River, in the suburbs of Moscow. Previously, I had tested preservation of pomegranate seeds in liquid nitrogen; seeds of many rare, threatened and vanishing plants have been preserved in liquid nitrogen. The pomegranate seeds showed excellent results. They were planted experimentally and their development was quite normal.

Preserving pomegranate pollen in liquid nitrogen was also tested and was found to be successful, too. After a set period of preservation, pomegranate pollen successfully grew in a sucrose solution both in a Petri dish and on pomegranate flower stigmas. When the resulting seedlings were planted, they grew quite normally.

My third conclusion from observing the peculiarities of inbred pomegranate seedlings relates to the practical collection of wild pomegranate seeds. Pomegranate sample collecting has its problems. You travel by air, by train, by car, on horseback. You walk miles and miles climbing mountains, crawling along gorges in the heat. Often it's dangerous and your ability to carry your finds back with you is limited. I almost always went by myself. Every new pomegranate sample included no less than 20 cuttings, each of them no less than 8"–10" long. As a rule, one sample weighed up to a kilogram. With seed samples the process is entirely different. You can easily carry many dozen seed samples back with you.

CHAPTER 14

# On Failures

**Among the first plant hunters** were the men Moses sent to assess the land of Canaan to report back whether it was rich or poor, forested or not. They brought back the fruit of the land. From that time on, we know that those of us who seek out and collect plants have our successes and our failures. There have been times when I was unable to find the plants or samples that were known to possess some unusual feature or a property valuable for selection. I knew of those plants from the literature or from descriptions but I couldn't find them.

I have already written about the blue pomegranate. That unique seedling was grown from seeds of a wild pomegranate that had been collected in the 1930s in the gorges of the southwestern Kopet Dag. As I've written, that seedling plant, together with the rest of the plants on A. D. Strebkova's selection site, were recently rooted up and ploughed over. Human stupidity and lack of foresight is limitless.

On all my expeditions, I spoke to local elders and gardeners about how their plants survived the 1920–1930 collectivization. During this time in the Soviet Union, well-to-do farmers were dispossessed of their land, cattle and property under an ideological program to create social equality under new collective management. Collectivization's bitter harvest touched my family, too. My grandfather in Byelorussia and my wife's grandfather in Tarhani in Russia had all their possessions taken. My other grandfather, Mark Levin, had no possessions at all so there was nothing to take from him. In Czarist times, an attempt had been made to permit Jews to own land. He had been allotted a piece of land not far from Odessa. He toiled on it, but twice there were droughts and everything he'd cultivated had dried up.

It always seemed possible to me that I might find ancient and rare pomegranate varieties in abandoned, pre-collectivization gardens that had been the pride of the local rulers in Central Asia. Alas, most of the time, the abandoned gardens had been destroyed, or not all of my samples collected from them grew well enough to become established in our collections. I'd have to return for samples again. More than once, I would learn that between my first and second attempts, the *bai's* (the local feudal landlord's) ancient garden from which I took my samples a year before had already been dug out by a bulldozer, ploughed over and seeded with cotton, the main crop of Soviet Central Asia.

I became obsessed with the idea of finding dwarf pomegranates in natural conditions. Long ago, dwarf pomegranate varieties had been discovered by chance on the Antilles Islands, in the Caribbean. I used that mutation as the starting point for my selection of an entire group of decorative dwarf pomegranates, including double-petaled ones of various colors. Attempts by my predecessors had failed, however, to transform that dwarf into a fruit-bearing variety. I was taken by that idea, too, so again and again I crossed the dwarf decorative and the best full-sized fruit-bearing varieties. My results were always negative, as had happened to the botanists before me.

Local shepherds, hunters and naturalists frequently would tell me where they had seen dwarf pomegranate bushes growing wild in the southwestern Kopet Dag. I was always very attentive and listened to their directions. Sometimes their advice got me where I wanted; sometimes it didn't result in my finding anything. When I was successful, I took cuttings from the dwarf pomegranate bushes that did not exceed a meter (three feet) in height, and collected fruit if there was any. The following spring, I planted the cuttings. After the plants were well established, we transplanted, irrigated and studied them in their new situation. Not a single one of those formerly dwarf bushes remained dwarf. The answer was that in their natural environments they overcame hard conditions and water shortage by being smaller. Their resulting dwarfness was a sample of phenotypic variability, where a plant's distinctive physical characteristics are determined by the interaction of its genetic make-up with environmental conditions. Dwarfness was a feature of environmental limitations.

In the 1930s, an article in *Soviet Subtropics* mentioned a pomegranate growing in the Ferghana Valley in Uzbekistan whose fruit had no separating membranes. Such a feature would be of interest for any horticulturalist, so in the early 1980s, I went to the Ferghana Valley and spoke with dozens of people, old gardeners and agronomists expert in local pomegranate varieties. All of them spread their hands. They couldn't help, and I couldn't trace that rare and valuable mutation.

No return was possible. The Ferghana Valley became an area of violence and bloodshed in the spring of 1989. Local nationalists fought against the Meschketin Turks whom Stalin had deported from Georgia to Uzbekistan during WW II. Georgia would not permit the Meschketinsky to return to their ancestral land, and they had to escape from Uzbekistan where they faced animosity. Fortunately, the USA offered permanent residency. These events were harbingers of the approaching collapse of the USSR.

I made an expedition to Kirghizstan—the former Soviet Republic that used to be known as Kirghizia—because I'd heard that wild pomegranates were to be found on the Ferghana ridge that bordered Uzbekistan. I had visited Tashkent, the Uzbek capital, to see the herbarium pages collected on the Ferghana ridge, but I knew no peace until I was able to verify for myself that sensational find in the pomegranate's most northeastern location of Soviet territory.

When I arrived in Kirghizstan's capital Bishkek (formerly

Frunze,) I contacted Professor Vasily Ivanovich Tkachenko, a prominent botanist. Tkachenko told me that he had been unable to find pomegranates on the Ferghana ridge, and that probably they had become extinct there. The region was far too cold for the pomegranate.

I decided to check the ridge area for myself. There were hardships reaching it, but eventually I came on a group of Leshoz foresters who managed the area. They listened to my story with surprise. They had never seen any pomegranates. Unfortunately, this wasn't the only time when a rare variety had disappeared from a certain location.

My friend Igor Nikolayevich Khlopin, an archeologist, wrote eleven books. Of the many he gave me as gifts, *What was there before the Flood?* particularly fascinated me. In it, he analyzed the Bible, the Book of Genesis, from the perspective of science. He showed that the biblical episodes and subjects were based on actual, prolonged historical processes that people had been unable to comprehend in their own time. I was most interested in the parts that described attempts to find the location of the Paradise on Earth.

The idea itself of Eden, the garden of the gods, was derived from ancient Sumaria. According to myths, the Sumarian paradise was located in a country called Dilmoon. The same place was believed to be the "country of the living" by the Babylonians who came after the Sumarians. In accordance with the Bible, Paradise was located in the country of Eden to the east of Palestine, which seemed to be this very Dilmoon. According to Khlopin, its location was in the Zagros Mountains in northern Mesopotamia and western Iran, an area rich with various edible plants.

I thought about this Eden when an unusual, I would say 'original,' man appeared at our station. He was short, dressed in a bizarre shirt, and he carried a guitar on a strap across his shoulder. He was a German named Joseph Gremlikh. Joseph had a rather good command of spoken Russian, and he had lived for years in the U.S.A., Brazil, Mexico and Finland. He also owned a farm in the Ukraine. He had a dream to turn the desert lands into a flourishing paradise. He said he was a modern idealist, and it was that idea that brought him to Turkmenistan to try to contact President Saparmurad Turkmenbashi, quite an unusual figure himself.

Joseph and I spoke at length about what location might be the best. I told him about Khlopin's book, about the paradise on the earth in the Zagros Mountains. I told him that I also wanted to visit that place from where, probably, many cultivated plants had originated, and about the wild pomegranate populations there. Joseph got excited about the idea of going. He promised to come back in the autumn with his truck to take me to the paradisiacal wilderness. Autumn came but Joseph did not arrive. Could there have been problems with his visa? I did not know. All I could say was that on Turkmenistan soil, Joseph's dream did not go any further. Nor did I arrive at his tempting paradise.

Many years before that time, I'd read in the newspapers that an expedition of Leningrad archeologists was working on the island of Socotra. My friend Khlopin helped me to contact the man who was leading that expedition. In my letter to him, I gave an outline of my interests on Socotra, where I knew four plants of the ancient *Socotria* pomegranate had survived. I wrote that it was of the greatest importance for science to describe *Socotria*. Would he collect seeds, etc? The expedition leader, whose name was Gryaznevich, amiably agreed to help me. Time passed, but I never received anything from the Leningrad archeologists. Just several days ago, on a nature program on TV, the anchor was speaking about his travels and showed the island of Socotra that he'd been lucky enough to visit. I saw its eerie ancient plants and felt very sad thinking about yet another of my dreams that was destined not to be realized, a dream that would have expanded our knowledge of the pomegranate family.

There are failures and then there are failures. Any interesting ideas I came across in popular science magazines, I mentally filtered for their applicability to pomegranates. At one point, I learned about a company in Estonia testing its chemical products as pheromones. Pheromones are organic compounds often secreted by the female of the species to attract the male from afar. The quantity of the pheromone may be extremely small. Synthetic pheromones are used in gluey traps to catch male insects, and in this way, to reduce insect reproduction. Each insect, however, has its own pheromone. The trap-makers' challenge was to synthesize and test numerous substances as pheromones.

I wrote a letter to the company in Estonia and suggested that they test their products on our pomegranates. The company responded, and for several years I tested their products, hoping to find the right one for *Plodozhorka*, an apple and pomegranate worm, the most common pest damaging pomegranate fruit. Some of the Estonian substances attracted apple worm males, stimulated by the smells these pheromes gave off, who would also eat pomegranates. But pomegranate-destructive males were stubborn in refusing to be attracted by any of the smells we used. The enterprise was quite time consuming. Every square hectare of our collection needed to have a certain number of insect traps, and they had to be refreshed every week. All insects had to be counted and described, along with whatever we were using to attract them. The company varied the compounds in their products annually. It was ironic that the traps attracted various insects, even mosquitoes, but not the males of our pomegranate-dwelling insect, *Plodozhorka*! It took me a while to cool down and let this project go.

Some time later, I read the work of an Azerbaijan researcher who had tested the same products with the same unsuccessful results.

After the collapse of the Soviet Union, I received two invitations to participate in international conferences on pomegranates. One was to take place in Spain, another in Tunisia. The new director of our station was a Turkmen, a specialist in irrigation who had only the vaguest ideas about our work on pomegranates. He said that he wanted to go to Spain himself, that both or neither of us would travel. I had to convince him to calmly accept the latter, with the pretext that neither of us were fluent enough in the working language of the conference.

As for the trip to Tunisia, an even more interesting detail surfaced that characterized the contemporary situation. The Arab countries denied entry to those who had ever previously visited Israel, and that included myself.

Thus, at the end of my scientific career, I was not able to speak about my achievements in punicology. My consolation was that a summary of my work was published in three articles in English by IPGRI, the International Institute of the Plant Genetic Resources in Rome.

In a Daghestan papacha

*Twilight*, Ann Getsinger

French silk tapestry, 1734

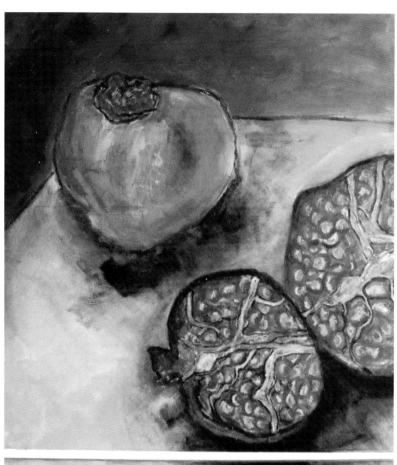

*Ecstasy*, Anitra Redlefsen

346. Punica Granatum L.          Granate.

Punica granatum

Bursting pomegranates

Black pomegranate

Flower and bud

Turkmen with pomegranates

Three mature pomegranates

DPUN 0143 "Sogdiana"    DPUN 0124 "Parfyanka"    DPUN "C...

UC Davis pomegranate tasting

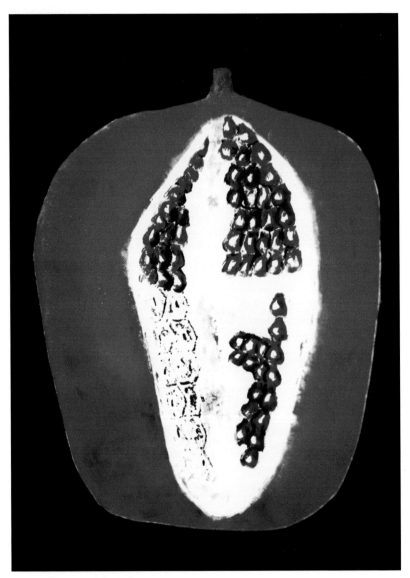

*Pomegranate III*, Celia Gilbert

# CHAPTER 15

# Botanical Mysteries

**A scientist is inevitably addicted to the unknown.** Nothing could be more true than when it's a matter of an addiction to pomegranates.

These days, detective stories fill up bookstores, TV screens, movies. I think that solving scientific riddles is no less exciting. A lifetime isn't enough for a scientist to understand a problem. I agree completely with a science fiction writer who said something to the effect that for an atheist, miracles are a scientific problem that probably cannot be resolved in the present but will certainly be in the near or distant future. Albert Einstein wrote that the most beautiful and profound human experience is the, "sensation of enigma, mystery." He believed that the unknown was the basis of all religions, art and science. All I can say is that the pomegranate contains so much that is unknown and unexplained that it remains the most tantalizing mystery for me. The same is true of many other plants and other biological subjects, of everything. Probably it will always be so. When one riddle is solved—or we believe it solved—dozens of others will surface.

I'd like to relate a personal experience about unclear phenomena that isn't related to the pomegranate but has remained with me as a fascinating enigma these many years.

The summer of 1941, I was in a young Pioneer summer camp in Tolmachevo near Leningrad. The Germans had just invaded

Russia. We learned about it four days later. My parents were mobilized to dig trenches around Leningrad so the children's stay at the summer camp was extended. They didn't change our regimen. After lunch we were expected to have a mid-day nap, the "dead hour." I, of course, disregarded it and escaped by climbing through the window. I wandered from the camp to the river. On the river bank, a tiny tree covered with *belyanka*, white butterflies with tattered wings, caught my attention. Their life cycle was ending and they had all landed on the tree to die. The tree was their cemetery.

That short episode made a profound impression on me. For decades afterward, I attempted to learn what the phenomenon was. I inquired, asked various specialists if they could explain more about the butterflies, but no one knew. From books and TV programs, I learned about elephant cemeteries, hippopotamus cemeteries, aquatic turtle cemeteries. I knew about the Monarch butterfly migration from Canada to Mexico that is similar to annual bird migrations. There are elephant migrations to the Himalayas in India that appear to have been explained by the changes in the air temperature and also related to elephant reproduction.

In 1961, during the first year of my long ordeal of post-graduate education, I worked from morning until night in the apple orchard. Only the unbearable mid-day heat could make me stop. There I saw multitudes of very dark blue dragonflies in the garden. For the next 40 years of my life in Turkmenistan, I never saw anything like them. What were they? Some waves of life? Was it a manifestation of long-term rhythms or cycles?

In the 1960s, I spent a significant amount of time in the Shikhim Dere Gorge where many of our subtropical plant collections were concentrated. A large part of the gorge hadn't yet been cultivated because of a water shortage for irrigation. The water we did have came from two little springs. At each spring, an enormous variety of butterflies clustered. They were so numerous that they attracted many amateurs from as far away as Moscow. Many times I saw those fanatics with butterfly nets rushing up and down the hills trying to catch some rare specimen.

At the end of the 1960s, a hydrologist from Turkmenistan discovered water in the top-most part of the gorge, in the so-called trace cone. A hole was bored, resulting in the release of seven gallons per second. The water had a slight smell of hydrogen sulphate but that evaporated fast. This tapped spring allowed us to cultivate large areas in the gorge and to add subtropical fruit varieties to our collections. Then, gradually, the butterflies disappeared from the gorge. Why? Were we to blame since we had intruded into the order that had existed in nature? Or was it the proverbial waves of life? Long-term rhythms, cycles? I can only guess.

In 1961, in the company of T. N. Ulyanova, I went on horseback to the summits of Sunt and Hasardag, the two highest mountains in the range named after them, about 5,900 feet above sea level. The Sunt-Hasardag separated the Sumbar River Valley from the northern Hodjakalin Valley that had a much colder climate.

The road gradually ascended to the Hasardag foothills. At the half-way point, on the left side of the path we saw strange, short gravestones, less than one meter high, and shaped as male and female figures. These made a powerful impression on me—the mountains that surrounded us, the valleys and the cemetery. It reminded me of Isaak Levitan's painting called "Over the Eternal Peace" with its rather oriental stylization.

The same year, on a trip to the Chandir River Valley, we saw a similar cemetery, but the gravestones—male and female figures—were much taller, three or four meters high. I had an inexplicable, bizarre sensation when I stood next to them. Some time later I read that M. E. Masson, an archeologist from Tashkent, had studied those ancient cemeteries. He thought they belonged to the pre-Islamic epoch, pre-sixth century A.D.

Once I went to southwestern Turkmenistan to study two pomegranate isolates—a distinct population determined by their location. In the Kizil-Atrek regional center we hired a local hunter as our guide. We started off before sunrise. In Aladag, our first destination located around 30 miles north of the Chat *kishlak*, among the boundless vastness of the Meshed-Messarian Plain, we found the two pomegranate groups. Just imagine that you are crossing a great plain that extends for dozens and dozens of miles, and suddenly you see a spot among enormous broken rocks that formed the trace cone, and next to bare limestone, you see seven pomegranate plants connected to each other. How did they get here? How had they survived through the centuries? The nearest wild pomegranate population in the southwestern Kopet Dag was in Songudag, 43 miles away. The distance was too long even for birds to carry pomegranate seeds.

Nine miles from there, we saw the second pomegranate group in the Cherkezli Gorge. There on the steep, almost perpendicular slope of blinding white shell rock grew six pomegranate bushes at a significant height, contrary to everything known about pomegranates. Nowhere else had I come across anything like them. It is hard to understand how the pomegranate could make a stand there, when and how the conditions were favorable for growth.

At this time, I thought about the pomegranates in relation to another event. On my way to the first bushes, in a shallow dip, like a small ravine or gully, there stood a vertical rock that was probably dug into the soil. There was an Arabic inscription carved on it. I photographed it and gave the picture to N. I. Khlopin, my friend the archeologist. He promised to show it to his colleagues who were Arabist historians. The origin remained a mystery, because the favor I asked was probably forgotten among my friend's busy daily chores and problems.

On our way to the white shell rock, we had passed by a circus-like sunken space with a diameter about six to nine miles that had high wall-like edges. I did not know what it could be. The crater made me think of a circus ring and I remembered it.

Later, I read about similar phenomena in other places on our planet and understood better what I had come upon. The majority of meteors that bombard the earth burn up in the atmosphere. Every year there are about ten entries of large explosions in the atmosphere. They are called astroblems. The power of the explosions can reach a megaton in some cases. About two thousand asteroids, known to have a diameter of more than a kilometer, are potentially dangerous for the earth. But even objects with diameters 55–108 yards can be dangerous. It's believed that large asteroids crashing into earth long ago could have caused the extinction of many large animals. According to one of the numerous theories on the subject, a gigantic catastrophe caused the extinction of dinosaurs. When I spoke with Turkmen geologists, they hadn't known about the astroblem that I'd found.

On another trip exploring for isolated wild pomegranate populations in the Chandir River Valley and the adjacent gorges, I went with a group of Moscow botanists. Frontier guards permitted us to cross the so-called control zone on a decent trail leading to the mountain range that separated Turkmenistan from neighboring Iran. Suddenly, on the underside of an overhang, we saw ancient rock paintings. These primordial artists had painted a hunting scene. In quite realistically depicted images, we saw primitive hunters with spears and the gazelles they were hunting. The paintings were good, but that wasn't the main thing. On one of the pictures there was a kangaroo! Everybody took their cameras to photograph the picture. Stricken by surprise, they began saying it was a sham, a joke, a falsification. As for me, I believe that the rock paintings in that isolated location were authentic and reached us from time immemorial. A very talented painter had done them. They were so expressive!

I cannot explain how the kangaroo happened to be in the painting, but I thought at that moment of a story told to me by one of my lab assistants whose nickname was "Soldier Aman." He was a serious man but he had insisted that in one of the gorges, people saw a huge snake with horns on its head!

Here I have an opening to introduce a little digression on snakes. Once, in the Upper Sumbar area, on a mid-spring day, I saw two large *ghurzas*, local poisonous vipers. They were standing on their tails opposite one another and rocking. It was a dance of two competing males during mating period. I prepared my camera. But as it clicked, the snakes immediately disappeared. Unfortunately I'd been in such a hurry that I'd forgotten to remove the lens cover. My companions and I told the episode to Victor Makeyev, an herpetologist. He did not believe us. He said that vipers like these *ghurzas* did not dance. It was typical of cobras but not *ghurzas* to dance. But I'd seen it with my own eyes! I protested. Makeyev was a Moscovite who used to come to Garrigala. Later, he wrote a little book on snakes and his own adventures, but he did not believe my account and I didn't have a picture.

Frequently, on my expeditions for pomegranates I saw other things that were quite unusual. Afterwards, when I began thinking about what I'd observed, I tried to find analogies, similar facts. I tried to get statistics, tried to find accounts of similar events. Time goes by sometimes a very long time passes before at least there's a preliminary explanation for the event that amazed me and occupied my thoughts for so long.

For example, the trunks of some trees for some reason spiral in one or another direction. This is the rather rare phenomenon of gnarliness. But in the higher areas of the southwestern Kopet Dag, I saw that phenomenon relatively frequently among many kinds of trees and bushes. I came across spiral gnarliness in pomegranates, too. I still do not know what caused the phenomenon. Nonetheless, I collected the data on spiral gnarliness in the area of high concentration and published a short article in a scientific journal.

The famous American dendrologist and popular writer, Edwin Menninger, called such trees "twisted." Twenty-two pages of his excellent book, *Quaint Trees*, that I read in the Russian translation, described the phenomenon and the opinions of various specialists. They had three different opinions. Twisting was caused by external factors in unusual conditions. A viral disease caused twisting. Twisting was determined by hereditary factors. Such uncertainty means that new observations and new research are needed.

In 1961, I was in Ghuyen Gorge in the direction of the Iranian border. The gorge reaches the Chandir River Valley in the southwestern Kopet Dag range. There, besides pomegranate bushes, we found a bush of *Mushmila germanica*, a Medlar, known botanically as *Mespilus germanica*. I recognized the species. When I lived in Daghestan in the Caucasus I'd come upon its wild forms on the banks of the Ghurgen-chai River.

That plant appeared to be an integral part of the Turkmenistan flora. But I had my doubts, since it was a single example. I suspected that the plant had gone wild, or rather, regressed into its wild form. That gorge had been previously the home of the Iranians. Some time later, L. E. Ishchenko, from the Ashgabat Botanical Garden, reported that she also found *Mushmila* in that gorge.

Many years passed. The Botanical Institute of the Turkmenistan Academy of Sciences launched an expedition to catalog rare plant species in the southwestern Kopet Dag. Edgar Seifulin, an energetic botanist, headed the group. He asked me to help find that *Mushmila*. I willingly agreed. Though many years had passed, I remembered quite well the small elevation next to blackberry bushes where the *Mushmila* bush grew. I explored around that small hill, but the *Mushmila* was not there any more. It was gone!

Border guards stopped our expedition from proceeding to the Iranian side of the border. A border guard told me he'd seen filberts there but I couldn't find them. The very fact that filberts were growing there supported my idea that *Mushmila* and filbert alike were once cultivated trees planted in the area, they had regressed into wild forms.

I remember coming across rare plant species in the southwestern Kopet Dag. I'm remembering first of all *Sinskya lucerne*—a species of alfalfa—in the Kurati Gorge. Orchids were tremendously rare in Turkmenistan but you could sometimes find *Listera oval* in the Yol-Dere Gorge, and *Fedchenko*, endemic orchids—a small tree with silver leaves—which was a new place for these orchid varieties.

I can say that the plant world is as inexhaustible as an atom.

Returning to the pomegranate, we must remember that we still can give only approximate and provisional explanations why, in many cases of seed reproduction, all a plant's features and properties are inherited without changes. As we saw earlier in connection with self-propagation, embryologists I spoke with were convinced that the pomegranate does not inbreed nor have capacity for self-pollination, and yet it happens. In inbred pomegranates, features and properties of the initial varieties are preserved in the progeny of the seedlings grown from the seeds that came from self-pollinated plants. Thus we can say that, no matter how much inbreeding occurs, certain identifiable traits are preserved and expressed in future generations. The more inbreeding, the more stable the traits become since they are genetically identical plants.

Here's another fact that's not ever been explained. In a number of cases, within the same pomegranate bush, trunks behave in individual ways. Some bear fruit while others do not. In other instances, all or only some bear fruit asynchronously, that is, not at the same time. Why is it so? I do not know. I would suggest that the phenomena relates to a certain genetic independence of the trunks. Whatever causes the phenomenon, it is comparatively rare for pomegranates.

Here is one more curious experience. In 2002, as my wife and I were leaving Turkmenistan forever—at least that's how it seemed to us as we departed from our home of 40 years—I took bulbs and seeds of some plants with me. Some I placed with the Institute of Applied Biological Research, at Ben-Gurion University in Beer-Sheva. Another part I kept for myself. In the fall, I planted bulbs of two types: Onion Supreme and Onion Paradoxical. I was especially interested in Onion Paradoxical, a small beautiful plant whose flower reminded me of a lily of the valley. I considered that plant to be a floral symbol of the southwestern Kopet Dag. The Turkmen call it *dere-sokon*, mountain onion, and eat its rather tender leaves in a variety of foods.

Strange things began to happen. In the spring, one or two bulbs of Onion Paradoxical grew and also one bulb of Onion Supreme, a very decorative plant in its native southwestern Kopet Dag region. Both were feeble and failed to reach average size. They stopped growing early. I thought they'd gone into their dormant state. The second year they did not grow at all, and I realized they had died. I was blaming myself for not providing the right care. Recently

Nadav Ravid, a consultant for Paramount Farms took me to Volcani Center in Bet-Dagan and introduced me to Rina Kamenetskaya, a specialist in bulb plant physiology, who gave me a more or less satisfactory explanation. She thinks that tulips and onions do not grow well in Israel due to the absence of necessary low temperatures for their normal dormant period. She probably was right.

I will not tire my readers by listing other unknown pomegranate and plant mysteries. There are many. What still remains mysterious to me among the ancient pomegranate varieties is their ability over years to pass on their features and properties at hybridization, what you might call their domination of hybridization.

Nature has many a *ghitiks*** many jokes and riddles.

---

** translators note: *ghitik* is a joke that refers to a trick with playing cards.

Allis Teegarden

# CHAPTER 16

# Pomegranates, Culture and Art

**In the beginning of the twenty-first century,** people live mainly in cities and relate to the world of plants only indirectly. We pass by greenery in urban parks, squares, boulevards. We put plants on our balconies or living rooms and aquaria. Woods and national parks make good weekend outings. For our educational benefit, we go to botanical gardens. We know that plants, especially forests, are the planet's lungs.

In the past, the plants played a far more significant role. In the beginning, our ancestors, hunters and gatherers, had a more utilitarian attitude towards plants because they needed them for survival. The relationship between humans and plants changed during the Neolithic revolution—the shift from gathering to growing what we ate. From the tenth to sixth millennia B. C., in various regions of the Near and Mid-East, those changes were in full swing.

In the process of domestication, plants became another phenomenon, taking on qualities they hadn't had in the wild. Unconsciously, aesthetic categories came about, and the beauty, splendor and magnificence of plants were represented in sacred categories. They became artifacts. Mythological thinking was an inevitable phase of developing culture, and myths about plants became a part of our cultural heritage. Many of these myths and their echoes still reverberate in festivities and celebrations honoring vegetation, in plant symbols, in the names that we give to countries, cities and streets.

The pomegranate, familiar even in pre-historic times, had a special place in the mythological world. Treasured both as fruit and as a medicinal plant, the ruby pomegranate became a part of the cults and arts of ancient civilizations of the East and the Mediterranean. The pomegranate appeared to have something unusual and peculiar about it, a mystery and an allure. In the entire history of the relationship between humans and the pomegranate, we find hundreds if not thousands of artifacts and passages in literature, affirming its worship. Were its seeds a coded message from the gods? Ancient peoples ascribed special powers to it and developed an ideal of beauty based on the fruit.

Over the years, I devoted a great deal of time to collecting pomegranate lore from various periods and people, studying and relishing the iconography and symbolism of this beloved fruit. The process appears to have had several phases. After domestication in various regions, the plant became an artifact. This stage has to do with cultivars and varieties. Last came the cultural phenomenon, when the fruit became a symbolic archetype, and mythic elements adhered to it. Later, as cultural traditions overlapped, we find pomegranate symbolism linking cultures in Asia, Europe and Africa.

As a botanist and student of the history of fruit, I believe that the apple of discord that Paris gave to beautiful Aphrodite on the Islands of the Blest had to have been a pomegranate. At that time, apple trees were not yet known in the Middle East. And I'm not the only one who believes the pomegranate was the seductive apple that tempted Eve into tempting Adam in the Garden of Eden.

"Lips sweet as pomegranate," was the image from the sixteenth century B. C. text from Ugarit (Ras-Shamra) in Northern Syria. In a twelfth century B. C. Egyptian papyrus preserved in the Turin Museum, a woman's beauty is compared to a pomegranate. We also have a poem from that ancient time, spoken as if by a pomegranate's voice: "The first of all trees am I! I would not be honored to be considered to be the second one!"

Nearly as early in history, we see similar imagery from the

ninth to sixth century B. C. in the *Song of Songs* by King Solomon, where we read, "Like a broken pomegranate are the cheeks of my beloved." In one of Aesop's fables from the sixth century B. C., a pomegranate and an apple tree are arguing over whose fruit is best.

In an old Italian fairy tale, a beautiful girl who came from a pomegranate fruit turned into a dove, and from a drop of her blood a pomegranate grew that cured many sick people. From the very last pomegranate, another young girl appeared.

In another story from India, a young and beautiful dancer whose name was Pomegranate asked the god Mahaprabhu to give her a husband and many children. The god agreed to grant her wish. She married a handsome man and had as many children as she had hairs on her head. Her husband, unable to cope with their progeny, ran away. Pomegranate, the mother, also attempted to flee but their children caught her and wouldn't let her go. She implored the god to save her from all her children. In answer, the god turned her into this very tree that lives for a very long time and bears a lot of fruit.

In Egypt, Hett Laws Second Table (16–18th centuries B. C.) commanded that a man found guilty of burning or felling a pomegranate tree must pay a significant fine amounting to the cost of six sheep. Pomegranate images were found in crypts of many Egyptian pharaohs. In Sumer in ancient Ur, in the crypt of Queen Shub-ad, (third millennium B. C.) the queen's head-dress is decorated with golden pomegranate fruits.

In Nimrud, Assyria, seventh century B. C., a small ivory pomegranate symbolizing fertility has been recovered. We know also that pomegranate blossoms were a favorite accessory among Babylonians, and the fruit was always served at wedding feasts.

Pomegranates symbolized fertility at celebrations honoring Mithras, god of light and truth in eastern Iranian regions, particularly in Bactria.

Pomegranates figured in Phrygean rites in western Anatolia. The fruits were a part of the cult of Cybele, the earth goddess, known in the myths from Parthia, Asia Minor and Central Asia as

Nana, Nanait, Anait, Anahit, and Atargatis. The nymph Nana was said to have been conceived from a pomegranate fruit, and that she bore Attis, who became a Phrygian deity of nature.

In Syrian-Phoenician religious ceremonies, the pomegranate was so significant that it was called "Rimmon," the name of the Sun god, "Gadad-Rimmon."

In Sogdiana, from the third to fifth centuries A. D., worshippers considered the pomegranate a symbol of fertility and abundance. Further east, it also symbolized love and fertility. Ripe and juicy pomegranate seeds were cast over the floor of a Chinese bridal chamber—the more scattered the better, expressing the hope for many children.

Saul, the first king of Israel, found shelter from the sun under a pomegranate tree.

In ancient Judea, King Solomon had a pomegranate orchard. Copper pillars and capitals ornamented with pomegranate fruit girded his temple. Pomegranate purple flowers and fruit were embroidered on the robes of Jewish priests. For ancient Semitic peoples, the multi-seeded fruit was one of the symbols of life and abundance.

Ancient Jews valued pomegranates so highly that when a tree perished it was considered to be a dire sign, an omen of Yahweh's displeasure. The pomegranate was a symbol of Jewish New Year, and its branches were used for Sukkot festivities, too.

In the Torah, the pomegranate is mentioned among the fruits that glorified the Land of Israel. Pomegranates are always included among fruits on the table to celebrate *Tu bishvat*, the New Year of Trees—a tradition that continues today in modern Israel.

The Koran refers to the pomegranate in many places. Prophet Mohammed indicated its high qualities by stating that he who eats it would get rid of envy and spite.

On Zanzibar Island, ancient original wood-carved doors always included a pomegranate as a symbol of fertility.

A pomegranate was a symbol of agreement, also a promise of faithfulness. Eleusinian mystery plays connected with the cult of Hades featured pomegranate seeds as a symbol and guarantee of spousal fidelity. Hades, god of the underworld, abode of the dead, agreed to let his wife Persephone visit her mother Demeter only after he made her swallow a pomegranate seed (or six seeds) assuring her return to his underworld. Since that time, Persephone spends two-thirds of the year in the surface world with her parents Zeus and Demeter, and one-third of the year with her husband in his underworld.

A pomegranate was often held up by Persephone and Demeter/ Ceres, goddesses connected with the land. In Mycenae, among Schliemann's finds, there was a small necklace that had several pendants in the shape of stylized pomegranates. Similar pendants were found in other places in Greece. On the island of Rhodes, an ancient image of a goddess of fertility has numerous nipples that look like pomegranate seeds.

In the *Odyssey*, Homer recounted that a pomegranate plant was growing in the orchard of Alkinoy, the Persian king. The ancient Greeks made pomegranate fruits of clay to be used in offerings to their gods. Today in Greece, an old custom has survived where scattered pomegranate seeds are believed to bring good luck in the coming year.

Between Argos and Mycenae, there is a temple with a sculpture by Polycletus portraying Hera holding a pomegranate in her hand to symbolize fertility. In the Olympian temple of Hera, on a Kipsela coffer, there is a relief image of Dionysus, where next to the god of drink and celebration, a pomegranate grows.

To express his passion, a youth in love would send a pome-granate blossom to a girl he loved. One legend recounts that on the island of Cyprus, Aphrodite planted a pomegranate in honor of her beloved Adonis after he was killed by a boar. According to another, Aerynias planted a pomegranate on the grave of Eteocles, and when fruits were plucked from it, the bush began bleeding. Pomegranate trees and fruit in Greek cults represented the chthonic mysteries of death, of fertility and rebirth.

Pomegranates were also associated with the torments of Tantalus, as a fruit hanging over his head, just out of reach of a man thirsting to death. The pomegranate was a symbol of both fiery passion and lasting love. To be deprived of it was torment.

A Persian legend tells of Farkhad who was in love with beautiful Shirin. When he killed himself, the very first pomegranate tree grew from his weapon.

In Persia in the late Susa period, a pomegranate twig was a required accessory to celebrate Nahvrus, New Year. A pomegranate twig is a part of a Parsee "Darun" cult where it is placed on one of the breads intended for ritual offering. In India, for the Durga Puja ceremony, nine plants are selected—the pomegranate among them.

Many peoples of the East held an ancient belief that only one seed in each pomegranate fruit had healing properties, and because they looked alike, you had to eat all of the seeds yourself, not lose any one of them, nor share the fruit with anybody.

Hellenistic images merged with the art of the ancient and medieval states of Central Asia (Merv, Samarkand, Termez, Khorezm) and Armenia, always associated with the idea of fertility. A Central Asian seventh century lantern decorated with a pomegranate tree and fruits was carried along the trade routes from Central Asia to the Russian area between the Kama and Viatka rivers. It was a symbol of hope for families with numerous descendants.

In Shibargan, Afghanistan, a gold pendant (first century A. D.) depicted a crown in the shape of a semi-naked woman holding a pomegranate in her hand. The motif of Anahit holding a pomegranate in her hand was quite popular in late antiquity and in early medieval sculptures and paintings (third to fourth centuries A. D.) We see the same motif—a woman with a pomegranate in her hand—in Indian sculptures as well.

In Siberia, between the cities of Tobolsk and Omsk, a silver medallion depicting Nike, goddess of Victory, with a pomegranate in her hand, showed there was a confluence of Central Asian and Greco-Bactrian artifacts in the region.

In some of the iconography, the pomegranate symbolized power. The shape of a pomegranate crown and cup inspired royal headdresses. There was a myth about Bacchus getting tired of his young lover and her desire to wear a crown. Bacchus turned the young woman into a pomegranate plant.

In Persia, the king's scepter was topped with a jeweled pomegranate.

In ancient Rome, pomegranate wreaths symbolized concordance, friendship and democracy. There was also the connection in Greek and Roman tombs, sepulchers and decorated sarcophagi, between the pomegranate and death, with the hope of rebirth that had always been part of the Persephone/Demeter myth. The so-called apotheosis type of tomb, with a portrait of the deceased in the center, often had a vessel with fruits placed under the human image. Sometimes there was a pomegranate sliced in such a way that its numerous seeds were exposed.

Even in Christian countries, the pomegranate had a special meaning as a plant that was mentioned in the Bible. In the Middle Ages, the pomegranate was associated with the Virgin Mary who had born an amazing fruit; pomegranate flowers became symbols of a death redeemed. In early Christian art, the image of a cut pomegranate was an emblem of hope for immortality.

A pomegranate branch with fruit on it was depicted in the oldest Armenian illustrated manuscript of the New Testament of 969 A.D. and preserved in Echmiadzin.

In Spain, pomegranate blossoms were the emblem of the Golden Age as well as of the city and province of Granada where it enjoyed preeminence in the Moorish kingdom. Pomegranates are still part of the Spanish national emblem.

Every fiftieth year in Chartres, France, the Cathedral Square is decorated with pomegranate blossoms.

The great Italian painters of the Renaissance picked up the pomegranate imagery. Rafael's first variant of the "Madonna Conestabile" depicted the Virgin Mary with a pomegranate in her hand. Elsewhere, the Virgin Mary was portrayed holding Jesus and a sliced pomegranate, showing that the fruit was still associated with resurrection.

Ancient coins with pomegranates from the fourth to sixth centuries have turned up in many places from the island of Milos in Greece, to Sydee in Pamphylia, Asia Minor.

Linguistically, we find the root words for pomegranate in the names of many towns, settlements and areas in ancient Palestine where the plants grew in abundance. In ancient Canaan, by the time the Israelites arrived, we find places named Rimmon, Gaf-Rimmon, En-Rimmon—all with the root word for pomegranates. The Moors named their most beautiful cities after the pomegranate. In Jordan, we find a city of Edad-Rimmon.

In Central Asia and in the Caucasus, among the Islamic peoples, I've discovered many proper names that include a generic Turkic root "nor, nar" meaning pomegranate. Normat, Normurad, Norkul, Gulnor, Anorgul, Nargul, Norbibi, Narinisoh, Norbeghim, Anortach are all names whose roots are from "nor" or "nar." We also find the name Anardara in Afghanistan.

*Geoponics*, the tenth century Byzantine encyclopedia of land cultivation, had a quotation from an African who repeated an ancient belief that a pomegranate branch was powerful enough to frighten away wild beasts. The advice was to place a branch at the entry to one's home. A Persian encyclopedia of the twelfth century had similar recommendations. Indians plant pomegranates in the center of their towns as a sign of respect for the plant's healing properties. Pomegranate wood was sacred for the Tajiks and Turkmen, and in the past was never used as fuel.

For pomegranate aficionados, there are more images and references than I can cite, from the frescoes of Pompeii where a rooster is pecking seeds in a sliced pomegranate, to the clay dish from Stari Krim (the Crimea) depicting a feast in a pomegranate orchard. On the sword handles of Xerxes' guard, there's a representation of the "pomegranate brigade" from the time of battle of Marathon (c. 490 B. C.); in the frescoes of Heraculanum; in the sculptures of Persepolis; in Buddhist cult centers from the second to fourth centuries A. D; in places like Kara-Tepe, Termez and Uzbekistan, where ornamental murals with pomegranate shoots and fruits have been uncovered.

We can still appreciate the silver vessel from the Urartu

era whose lid has a gold pomegranate on the top; the frescoes of Afrasiab (Samarkand, third to fifth centuries B. C.) and Penjikent (ninth century A. D.); a clay ossuary (Uzbekistan, seventh to eighth centuries A. D.); on the column capitals of a Zvartnoz temple in Armenia (seventh century A. D.); painted decorations of Shekhin Khans palace in Azerbaijan (eighteenth century A. D.); in the portals of the houses in Granada, Spain, and on Chinese porcelain dishes of the eighteenth century.

The popularity of pomegranate images on Indian seals, Persian rugs, Chinese silk and porcelain goods, velvets from Florence and Spain, tapestry from France and Flanders and brocades from Turkey, continues today in fine and decorative art. I've found sixteenth century English wallpaper with a pomegranate design. Salvador Dali painted his "A Second before Waking from Buzzing of a Bee flying around a Pomegranate."

Curiously, other plants carry the pomegranate adjective: pomegranate cherries, pomegranate grapes and pomegranate grapefruit.

I continually come across pomegranate motifs in paintings and literature created by Soviet artists. In Uzbekistan, A. N. Volkov painted his "Pomegranate Tea-House." Moscow artist A. I. Kravchuk placed the pomegranate against the background of Sunt Mountain after a visit to Garrigala. When my wife and I left Turkmenistan forever, we sent Kravchuk's still life to our daughter in Moscow as a memory of her birthplace.

Far away as I am from my former homeland, I've been carried back by reading references to pomegranates in Mikhail Bulgakov's *Master and Margarita*; Olga Forsh' *Last Rose*; Fazil Iskander's *Oh Marat*; Gergy Leonidze's *The Tree of Desire*. The pomegranate made an appearance in the Azerbaijan drama *Pomegranates in Our Settlement* by Rahman Alizade; in films by Sergey Paradjanov, *Ashik-Kerib* and *Pomegranate Blossom*; in T. Abuladze's *The Tree of Desire*"; in the Azerbaijan film, *Memories of a Pomegranate Tree*; in the Uzbek cartoon, *A Tale of a Fairy Pomegranate*. And last but not least, the Armenian-American William Saroyan called his work, *The Return to the Pomegranate Trees*.

Nowadays, in my home in Israel, as one eats pomegranates on the Jewish New Year, one says, "May it be your design, oh Almighty, that our achievements multiply like pomegranate seeds."

# CHAPTER 17

# Pomegranate Cuisine and Wine-making

**The pomegranate was introduced into culture** about 5,000 years ago. We can find the very first recipe using pomegranate seeds in an ancient Egyptian papyrus. From then on, the ruby globe enhanced and decorated dishes in the Caucasus and the East. The West has been slower to catch on to the sweet-tart flavor and bright colors, but that's changing.

I've estimated that at the present, the pomegranate ranks seventeenth in global fruit production. Its annual harvest may be from 800,000–1,000,000 tons. For some unclear reason, IPGRI (Rome, Italy) has placed pomegranates among plants of lesser distribution. I do not think that's correct, especially if you consider pomegranate distribution in the regions of Muslim culture. With contemporary technologies, growing pomegranates can be quite profitable, especially if juice production begins to tap the wider global market.

The pomegranate's juice is used for grenadine, compotes, syrups, wines, liqueurs, aperitifs, punches, jellies, jams. Pomegranate *narsharab* is made by slow, long cooking of sour wild pomegranate fruits and is served with meat or fish dishes. *Nardancha pastila* combines pomegranate and grape juice and adds delicious flavor to ethnic dishes. During one of my expeditions in Armenia, I visited a winery that used a simplified method of pomegranate winemaking by squeezing the entire fruit. Traditional methods, however, separate pomegranate seeds from the rind to prevent the juice from being too astringent, as high astringency makes poor quality wine.

My wife Emma and I learned over the years how to use our pomegranate crop from our garden. When you work under conditions peculiar to your own kitchen, you develop know-how the way Emma did. Each year, Emma's pomegranate juice improved. Of course, it depended on which varieties did best in any given year, and on the maturity of the fruit, how much sugar they'd stored and how much anthocyanin was present—upon which the deep red color depends.

The technology of preparing pomegranate juice at home isn't complex, but you must follow certain steps. You first extract the grains, which we did manually. We pressed the juice, also by hand. We found it best to squeeze the juice, without using any metal appliances, through a soft dense fabric or doubled fabric of a chintz type sewn up into a bag. We then kept the juice for at least 12 hours in enamel containers in the refrigerator so it wouldn't begin to ferment. We carefully poured juice from our decanters into wide-mouthed glass canning jars, filling them almost to the top before we sealed the jars with rubber-lined lids or screw-on metal covers and put them in a water bath that covered the jars. We heated the jars on our gas stove just to the point before the juice would have boiled over, approximately 158–160°F. We then removed the jars, screwed the lids as tightly as possible and turned them over on something soft. We let them cool and had many containers of garnet red juice that could be stored for up to half a year.

When we had lots of juice we made wine. There were two variations. For dry wine, we poured juice into large bottles and fastened the neck with gauze so there would be an exchange of air with the liquid. Fermentation went on in our bottles. Any overflow went into another bottle. When fermentation was over, the wine was ready. Mature fruit with a high sugar content made a pleasant, gentle, semisweet wine, but it wasn't stable. In our heat, bacteria could grow and the wine would quickly deteriorate.

Our problem was how to prevent this damage in a hot climate.

We began to prepare pomegranate liqueur from the wine. We added sugar, about a third of the volume of the juice, attached a glass cylinder with gauze—again for access of air. The sugar gradually dissolved and fermentation occurred slowly. We judged our success by how long the liqueur lasted—sometimes for many years.

During my expeditions across Central Asia and Trans-Caucasia, I came across various uses of pomegranates. Thinking back on the magnificent Georgian cuisine that cannot do without the pomegranate, I came home eager for Emma to try recreating the dishes, and she did so. We added the seeds to our salads, but as we didn't yet have soft-seeded varieties, ours were hard to chew.

In Tajikistan, I learned that German colonists cooked jam from pomegranate seeds, but when Emma tried, the jam wasn't quite right. There must have been a secret to the preparation that we didn't have. Also in Tajikistan, I found a local population that made a popular pomegranate wine called "mussalas." It was prepared in small amounts for the important people. I don't know what their methods were.

In Uzbekistan at the agricultural station, I was treated to a "grenadier"—a mix of pomegranate juice and spirits. This drink didn't give me much pleasure.

Often in autumn when I went along the Sumbar River toward the heights of the mountain range, I noticed that on the roofs of houses there was a layer of pink. The Turkmen were drying grains of a wild sour pomegranate and making what they call "turshi" (sour), which they added to their noodle dishes.

# CHAPTER 18

# The Pomegranate as a Cure for All Ills

**The pomegranate is a poly-vitamin,** a unique machine producing a wide spectrum of biologically active substances—ever more important for health in our polluted environment. Pomegranates are good for getting rid of waste from organisms including radioactive substances. It is an antioxidant, a tonic and a restorative, prescribed after infectious diseases, surgeries or as a general stimulant.

In the USSR, cosmonauts, pilots, submariners and coal miners were given pomegranate juice for their general health and for stamina. Monkeys on biological research satellites were fed pomegranate-rosehip juice for vitamins.

The pomegranate as a cure-all is nothing new. In ancient and medieval pharmacological manuscripts, it was valued as a powerful medicinal plant, used in folk medicines from Mediterranean countries to the East, and in Africa as well. Pliny, the Roman historian, considered the pomegranate a universal medicine. In our time, University of Oslo researchers include pomegranates in their list of "long life products."

I believe that ancient knowledge of how to use medicinal plants should be a part of contemporary scientific data and our personal health regimes. If you collect folk remedies as I have, you'll see pomegranates used for just about everything: asthma, coughs, colds, laryngitis, bronchitis, respiratory infections, weight problems and diabetes, headaches, brain diseases, heart problems, malaria, fever, colitis, hemorrhoids, colon problems, scurvy, jaundice, liver, kidney, spleen and gall bladder diseases.

An ancient Egyptian medical papyrus, probably dating from the ninth year of Pharaoh Amenhotep's rule in the sixteenth century B.C., gave a compilation of data about medicinal plants that included the pomegranate among many remedies.

Hippocrates (460–377 B. C.) recommended pomegranate juice for stomach pain, and pomegranate rind for ulcers. The pomegranate's medicinal qualities were highly appreciated by Theophratus (c.372–287 B. C.) who was the father of botany; by the doctors Galen (c.130–200 A. D.), Dioscuri (first century B. C.), Oribatius (325–403 A. D.), Paul of Aegean (seventh century A. D.), Johann Damascean (ninth century), Er Razi (850–932), Johann ibn-Masua (eighth to ninth century) and Mhitar Gheratsi (twelfth century.)

In his *Canon of Medical Science*, Avicenna (980–1037) mentioned pomegranates 150 times, considering all parts of the plant to be medicinal.

Quatrains in the 1512 *Rubaiyat* praise the pomegranate's medicinal qualities. "Punicotherapy" (from *Punica*, Latin for pomegranate) is a familiar term in some parts of the world, especially in the countries of Islam and Buddhism. Some South Asian countries grow pomegranates as special medicinal raw material.

For many centuries pomegranate rind as well as the bark of the branches were used to get rid of helminthes (intestinal parasite worms.) Its rind was used to treat stomach disorders, diarrhea and dysentery. Korean folk medicine employed it to treat severe dermatitis.

In various folk cultures, pomegranate fruit and other parts of the plant are used for stimulating appetite and to treat other stomach disorders like dyspepsia, gas, nausea, regurgitation, hiccups, to guard against spasms, dropsy, ovarian problems and shortness of breath. It is good as a painkiller. In India it is believed to be effective against leprosy. Indian folk medicine uses a pomegranate tonic for the nervous system.

During WW II, Azerbaijan produced lemon acidic sodium from wild pomegranates. In 1942, it was the best known blood preservative and so was sent to Leningrad during the siege. That same lemon–citric acid in pomegranate juice is used in treatment for scurvy and uric acid diatheses.

Pomegranate juice is good in treating arteriosclerosis and hypertension, and is well known as an excellent treatment for anemia.

You'll find references to pomegranate treatment for kidney stones, arthritis, night-blindness, baldness, eczema and skin problems of many kinds, fractures, hernia, small pox, leprosy, burns, malignant tumors and infertility.

We find the pomegranate in love potions as an aphrodisiac and for virility. Pomegranate seeds' rich oil has hormone producing effects and stimulates estrogen production. If you relied on folk medicine in the past and needed a tranquilizer for nervous disorders, or for snake and scorpion bites, you probably would have been prescribed some part of the pomegranate.

Contemporary phytotherapy, the science that studies the usage of plants for a wide spectrum of ailments, has a central place for the pomegranate. The rind has a suppressant effect on abdominal typhus bacteria, TB, intestinal dysentery, cholera and other bacteria that cause amoebic dysentery. Every part of the plant is being experimented with for potential benefits.

Pomegranate fruit has a high content of riboflavin—the B2 vitamin that normalizes the nervous system and is used against radiation sickness. It contains folic acid, steroidal estrogens, polyphenols, anthocyanins (having high capillary-strengthening activity) coumarin to treat hypertension. Pomegranates have an anti-ulcer, anti-coagulating, pain-killing effect, and contain an adrenalin-like bactericide. The fruit contains oxycoumarine (valuable in preventing strokes, thrombosis, clots, fractures in blood vessels,) betatine, an anti-ulcer organic compound, tocopherol or Vitamin E that has anti-radiation and anti-mutation activity; phenolcarbon acids, amino acids, pectin, small amounts of two alkaloids, 17 micro-elements including potassium, calcium, magnesium, molybdenum, copper, iron, cobalt, chrome and selenium.

Pomegranates contain arachidonic acid, an essential polyunsaturated fatty acid very rare in plants, and protocatechuic acid, both of which participate in prostaglandine synthesis, and consequently, influence the reproductive system. The leaves have tannins that are being tried for antibacterial and capillary-strengthening effects. Decoctions made of pomegranate flowers, fruit and roots may find their usages as well.

Most contemporary research on the pomegranate's pharmacological and medical benefits are being carried out in Israel. We may see parts of the pomegranate used in estrogen replacement therapy, to fight breast and prostate cancer, as well as inhibiting HIV infection.

# CHAPTER 19

# More Uses for the Pomegranate

**I remember an old man in Garrigala** mailing numerous boxes of dry, wild sour pomegranate rinds to Tashauz in the north of Turkmenistan, where they would be used to tan and dye sheepskins for the traditional *tulup* coats.

Oil from pomegranate seeds has excellent taste, but it's also been used for producing long-lasting glossy emulsion paints.

Pomegranate wood is a hard timber used for carving small objects. In Jericho, Palestine, remnants of cups and vessels made from pomegranate wood have been excavated from layers of the mid-Bronze Age (1650 years B. C.)

Pomegranates are valuable nectar containers, and very attractive to bees. One hectare, about two-and-a-half acres, of pomegranates in an orchard yields more honey than a plum orchard. Only apple and cherry trees provide more honey.

The pomegranate is a good choice for reforesting mountain slopes and has been used to ameliorate erosion.

From Babylon's hanging gardens to orchards and parks, in cities and palaces in the Near East and Central Asia, pomegranates have long been prized for the beauty of their bright leaves, blossoms and fruit. The pleasure they give lasts a long time, sometimes 10 months a year, starting with the first buds, through blossoming into fruits. After the leaves have turned yellow and fallen, fruit remains on the branches, ranging from bright red like Chinese lanterns, to dark purple, almost black.

Decorative pomegranate varieties are shaped as little trees with weeping twigs or as bushes combining several plants with flowers of different colors. Blossoming on decorative types lasts two to three months. Decorative plants can be grown in gardens, parks, squares and as border plants along roads. They make beautiful cut flowers.

Dwarf long-blossoming varieties are grown in pots for early flowering. They may have double-petaled flowers in red, white, pink or cream. The Japanese use dwarf pomegranates for bonsai culture.

Pomegranate plants tolerate pruning and trimming, so they are widely used as hedges. The plants don't require rich soils. They tolerate wind, smoke, dust, car exhaust and industrial pollution, thus making them a promising plant to grow along freeways and in industrial centers.

Expedition with Turkmen

Turkmen family

# CHAPTER 20

# THE SUMBAR RIVER VALLEY

**You already know about my good friend,** the archeologist Igor N. Khlopin. He wrote many scholarly books and articles, and we collaborated on pieces describing the Sumbar River region for local Turkmenistan newspapers. Much has happened in the decade that has passed since we worked together. The Soviet Union has collapsed. Our agricultural institute at Garrigala has been placed under the control of the Turkmenistan Ministry of Agriculture. In 2000, Garrigala celebrated 70 years as an agricultural station, but the anniversary was not a happy one. The station was deteriorating from lack of administrative attention and lack of funds. The most experienced researchers had died or left. Plant collections were perishing. Collections that had no equal in Central Asia nor the world, all were being lost. The new administration in Ashgabat did not care.

Igor N. Khlopin died in 1994. We had spent half our lifetimes researching the Sumbar Valley region. Witnessing the changes in the valley and at our station became unbearably hard. So I left, saying goodbye to the mountains and to the Sumbar Valley. I cherish a belief that our work was not in vain, that the valley will remain a magnet for researchers, nature-lovers and tourists.

The southwestern Kopet Dag mountain range, part of the Turkmen-Khorasan system, lies in the southwest of Turkmenistan.

It is mainly a fold formation from the Pliocene Quaternary period. Its value for us as a natural laboratory for studying the entire Turkmenistan and Central Asia territory is truly great.

A visitor arriving for the first time in the southwestern Kopet Dag on the way to the Sumbar River Valley is struck by its moonscape badlands—a landscape unlike any other in the entire territory of the former Soviet Union. The road that descends to the valley near the Garrigala settlement—formerly Alexandrovka, founded in 1892—is the center of the Garrigala region.

Two brooks, the Dainesu and the Kulun-Kalaci-su, merge near the Daine settlement and become the Sumbar, a moderate size river 152 miles long. The Turkmenistan part of its basin occupies 4,500 square miles. Two mountain ranges protect its valley from the wintry, icy breath of the Arctic that sometimes reaches the Kara-Kum Desert and the lowland plateau of the Kopet Dag. Nature in the valley of the river as well as its tributaries and the adjoining gorges is noticeably different from that found in the surrounding mountains and deserts. It is subtropical.

In its middle region (984–1,968 feet above sea level) the Sumbar River Valley is open to the west. The valley's climate actually has both subtropical and arid features; its mean annual air temperature is 61° F, equal to the mean temperature of the earth. The valley's temperature in January is 38.5° F; in July, 84° F; absolute minimum is 3.5° F. Generally there are 236 above-freezing days annually and a yearly precipitation of 13 inches.

The favorable subtropical climate has triggered the manifestation of valuable and unusual qualities in plants. In this unique natural laboratory, many phenomena could be observed that aren't found in other environments in Turkmenistan or elsewhere. Many trees and briars manifest neoteny, the unusual development of blossoms or fruits in their first year, as well as multiple annual blossomings and fruit bearing with other anomalies in growth and development. In local greenhouses, cotton varieties from several continents bloom all year.

In the very center of the Garrigala settlement stands a monument to Magtimkuli, a great Turkmen poet and thinker. Twelve-and-a-half miles to the east of Garrigala, in Uvankala, there is a Magtimkuli museum dedicated to the poet, and also to Alayar Kurbanov, a Civil War hero.

Zelili, Magtimkuli's father, himself a famous Turkmen poet, was born here. Not far from the settlement, in the Parhai Gorge, you find a thermal hydrogen-sulfur spring with water rich in micro-elements, and probably also radon as well. Only local people and a few visitors have used the spring.

Not far from Garrigala, in the Bagandir Gorge, a serpentarium uses another one of the region's natural riches: its serpents produce over two pounds of venom annually—a valuable raw product for the pharmaceutical industry.

The middle run of the Sumbar River has a rather wide valley. The *adirs* (lowland hills) and the mountains are to the north and south. In spring, the slopes are green, while in summer the grass fades and burns, making the landscape look lifeless. In the gorges, however, vegetation remains green into autumn. The Sumbar River itself almost dries up in the summer, but if there's rain in the upper river area or in neighboring Iran, the water level in the river rises and roaring mountain torrents rush down past the villages, carrying away the soil, uprooting trees, and causing damage.

In recent years, such mountain torrents have happened more frequently—the consequence of felling mountain forests and over-using mountain pastures, which increases soil erosion—everything we call "anthropogenic pressure." As recently as the beginning of the twentieth century, the Sumbar River Valley was covered with thickets and *tugai* forests on both banks. Nowadays, many of the slopes are bald. Only a short part of the lower Sumbar River still has some remnants of *tugais* along it. During the last 70 years, the water level of the springs, brooks and rivers has shrunk to half of what it used to be.

These days, in the Sumbar's middle course, the valley presents many miles of almost uninterrupted pomegranate gardens. Above Ai-Dere Gorge, gardens of cherries, apricots, apples, pears and other fruits are flourishing.

Since the 1870s, when Russia colonized the Trans-Caspian Krai (present-day Turkmenistan), many researchers were attracted by the astonishing natural treasures of the southwestern Kopet Dag—a region that contrasts sharply with the lowland Kopet Dag and the Kara-Kum desert. Many of the foremost botanists were profoundly impressed by the richness and uniqueness of the flora in the gorges and on the mountain slopes.

In May, 1912, B. I. Lipski, a famous Russian botanist, and A. I. Mihelson, a collector, departed from the Garrigala settlement heading to Yol-Dere (Road Gorge). They began ascending Sunt Mountain early in the morning. The summit, covered with dense woods, appeared not too far away, but once the botanists were deep in woods lying before it, they understood that they would not be able to penetrate them that day. Later Lipski wrote: "The woods were all over the cliffs, or rather, the woods have grown among many dense cliffs. The trees were not very large but they grew so densely that it was very difficult to get through them. The woods reminded me of a tropical forest; as in those regions, the trees had lichen growth." Fifty years later, I took the same road many times—if one can call it a road—to the mountain top.

In M. G. Popov's account of a subsequent exploratory trip, he wrote: "Nowhere in Central Asia and Turkmenistan is there the richness and variety one finds in the woods of the Western Kopet Dag."

Academician Nikolai Vavilov visited this distinctive area in Central Asia several times between 1916–1936. On a train to Ashgabat, after visiting the agricultural station, Vavilov wrote that the Garrigala area was one of the world's richest subtropical sources for priceless fruit varieties, with a concentrated, sizeable number of species and sub-species.

There are 42 varieties of fruiting plants in the gorges. Wild pomegranates, figs, grapes, cherry-plums, blackberries, hawthorns, apples, pears have their origin here. All in all there are 720 kinds of useful plants in the southwestern Kopet Dag, among them 311 that are medicinal, 142 that are decorative, and 131 that are volatile and oil-bearing.

The unique location of the southwestern Kopet Dag at the junction of the Mediterranean area and Turan lowlands, the northern outpost of the Iranian tableland, have determined its complex biological history and the extreme peculiarity of vegetation. Flora of this region is characterized by very high endemism, or uniqueness to its particular locality. Eighteen percent of the varieties of plants typical to the entire Kopet Dag are endemic, found only in the region and nowhere else in the world. Among lower plants, of 417 kinds of algae, 57 are typical only for the region. *Turkmen mandragora*, endemic to the Sumbar River basin, is an amazing plant with many legends surrounding it. Mandragora is endangered and listed in the Red Book of protected plants in Turkmenistan and the former Soviet Union.

Vavilov wrote that this special region, the origin of so many plant species, was a priceless treasure in need of protection. In 1930, Vavilov established the Garrigala settlement as the home for the Turkmenistan Experimental Station for Plant Genetic Resources. Seventy years of intensive work, expeditions to collect plants, as well as exchanges with other research institutions, resulted in a collection of 4,250 varieties of grapes; southern and subtropical fruit and nut cultures including olives, figs, persimmons, apricots, jujubes, and the largest pomegranate collection in the world with 1,117 accessions, or individual living plants of different varieties.

Over the years, employees of the agricultural station, and particularly O. F. Mizgireva, its director, spent many years trying to convince the public and the government of the urgent need to create a national park, a *zapovednik*, in the southwestern Kopet Dag. At last, in 1973, a smaller version of the envisioned protected area became Sunt-Khasardag Zapovednik, covering 74,131 acres in the basin of the Sumbar River. Its fate was both typical and pathetic—this reduced territory was not protected by a buffer zone. As years went by, the territory was still further reduced and a game reserve was allowed, legalizing pasturage of cattle on park land—a far cry from preserving ecosystems of flora and fauna.

In Turkmenistan, the national parks occupy only 2.2% of the entire territory, while they should make no less than 6% to provide for the protection of its flora and fauna, its rare biocomplexes and landscapes. The area of Sunt-Khasardag Zapovednik makes up less than 2% of the Turkmen part of the Sumbar basin.

Three-quarters of the way through the twentieth century, 5% of the plant varieties had vanished in the Sumbar basin. *Archa Turkmensoy*, local juniper, suffered most of all. Overgrazing the mountain steppes has caused a great deal of damage.

In 1990, a scientific study predicted that in 10 years, Garrigala soil erosion would increase 1.5-fold; in 20 years, 2.5-fold. Wooded areas would also be reduced along with good pasture areas. The scientists predicted that even if the Garrigala area became a national park, there wouldn't be noticeable improvement in the near future because of the extent of the destruction and the slow natural processes of recovery.

The fauna is also rich in the Sumbar River Valley. Seventy-six kinds of mammals are found here, 80% of the total in Turkmenistan. 238 kinds of birds are left, six of them protected and listed in the Red Book of Turkmenistan. There's a rich world of reptiles here—41 species, nine of them protected. Among insects, there are 102 species of *orthopterae*. The Sumbar Valley has 79 kinds of ants—19% of them are endemic, found only in this region.

Fifty years or so ago, in the Sumbar Valley, there were Turan tigers, cheetahs, Kopet Dag bears. They no longer exist. Leopards, hyenas and honey-eaters, as well as many others, are now rare. It is not easy to preserve them in the Sumbar basin with the increasing human pressure and the current tendency of the Turkmenistan government to reduce the territory of the national parks.

The rich valley of the Sumbar River was populated long ago. Archeologists determined that the traces and remains of Stone Age man here go back to 100–300 millennia B. C. Contemporary data has revealed a vast portrait of humanoids. It's been suggested that the territory of present day Turkmenistan should not be excluded as a probable site for the origin of humans.

Archeology in the entire Sumbar basin by Russian Academy of Sciences expeditions uncovered burial grounds from Neolithic and Bronze Age periods, from the end of the fifth to the end of the second millennia B. C., unique not only to Central Asia but to the entire Near East. From these finds, there has been a new interpretation of the peoples who populated this region. The area, called Ghirkhania (or Goorgan, Gurgian) has long been inhabited.

Scientists have shown that the *Avesta*, the sacred books of ancient Iranians, have their origins in the basin of the Tegen and Murgab rivers, now East Turkmenistan. The exception is the Gahts portion that the prophet Zoroaster recited during the mid-sixth century B. C. The *Avesta* goes back to the culture of the Bronze Age when polytheistic cultures developed in settled agricultural oases, before cattle-breeding tribes entered Central Asia. The Aryans, who created the *Avesta*, described only Aryan lands and countries, though other Iranian-speaking peoples (foremost the Toors, enemies of the Aryans for many centuries) are mentioned. It appears that the Aryans and the Toors lived in adjacent territories.

In a number of historical and geographical works of antiquity, the Tegden River in southeastern Turkmenistan was called the Arius. The Aryans lived in the valley of that river and had many cities there, including their capital called Artakoana, subsequently destroyed in 329 B. C. by the armies of Alexander the Great.

The Sumbar Valley is the only place on earth where the culture of the ancient Iranians retained its original images without contamination from foreign incursions that were typical in the history of Indo-European peoples. Further archeological studies of the Sumbar Valley may lead to new discoveries about those amazing times and peoples.

The Sumbar River basin is rare for all it offers in natural, cultural, historic conditions. One cannot wait to preserve it until human impact, as well as desertification, has gone too far. The area of protected Sumbar territory needs to be expanded four-fold and should be made a national park. The Sunt-Khasardag Biosphere

Park should be made its core. As it is now, the Sumbar Valley continues to be severely damaged despite efforts by environmentalists to preserve it. The valley must be declared a World Heritage Site by UNESCO. If protected internationally, then the Sumbar Valley region has a chance to retain its uniqueness.

# CHAPTER 21

# Long Routes with Unusual Transportation

**For 30 years, I made annual expeditions** across Central Asia and the Trans-Caucasus. As a rule, one expedition was to the left or west, to Trans-Caucasian countries, and another one was to the right and east, to Central Asia.

Preparations were easy when I was going around the south-western Kopet Dag region. The distances were not great, and if I needed to return, I could always do so. Difficulties arose when I had to travel to the other republics by air, train, bus and then on foot, frequently without a guide.

I would read literature on the region, locate where pomegranate plants were growing, whether they had been researched or studied, and to what degree. I learned everything I could about getting there and contacted the researchers who specialized in the region. Then came the many bureaucratic hurdles. Sometimes I'd ask for two trips to a distant region so that I could initially acquaint myself with the wild pomegranates or old gardens during the blossoming period, and then make a return trip.

In the cultivated zones, the old gardeners were invaluable. You could find them in practically every village, every settlement. On my paths and roads, I met many old men who became unselfish assistants once they understood our goals and problems. I'd talk with them, ask if they knew of any old gardens from the times before the 1917 Revolution, and later, the collectivization had completely destroyed their traditional way of life. Frequently the old gardens were no more than a pathetic and abandoned site with neglected bushes that bore no fruit; some didn't have a single shoot, not even enough for a cutting. Sometimes, I'd cut such bushes severely on my first visit so they'd send out new shoots the next spring when I returned.

Some of these old men were truly memorable. In Karakel ("Black Baldness") a village in the Chandir River Gorge, I remember a man whose nickname was "Goshi" that meant "flipper" because his fingers on one hand had grown together. Goshi cared for an old garden, maintaining it in rather good order. I was told that an Arab from Iran had started the garden with good varieties of fruit trees as well as grapes. Gradually, both fruit trees and grapes from that garden had been distributed throughout the area.

Researching wild pomegranates could be quite complicated and unpredictable. First of all, in wild populations, plants don't blossom and bear fruit every year. What happens depends on the condition of the plants and the dates of the most recent severe winter storms that may have done damage, or alternatively, a recent drought. All too often, a wild pomegranate population that people knew existed had been destroyed before I reached it. I always felt that I worked under a time constraint; I had to decide whether to stay long in the same location or explore as many locations as possible. I was never sure if I would be able to return.

I remember our last deputy manager responsible for planning expeditions used to say, "Make your expeditions while they are paid for!" He was right. That moment came when the Soviet Union fell apart and our VIR expeditions ended.

Looking back, I can say that the best means of transportation was what was available to me. I had little to choose from. Anything that might take me to my location was what I used. Often that meant walking. You needed endurance and stamina to be a pomegranate researcher. If you did not have stamina and persistence initially, you had to develop it. I started training early.

My first expedition in 1950 was to Kabarda in the North Caucasus, collecting live plants for the botanical gardens of Moscow University. I was at the head of a mule caravan on the border of Svanetia, one of the highest elevations in Georgia. We had to go through two passes in the Adir-Su Gorge—a long way to walk—plus I had to manage our four loaded mules. They gave me a lot of trouble. I'd had no previous experience with these animals, but I learned on the way.

I'd noticed previously in Kabarda that local hunters and shepherds all wore special footwear for walking in the mountains. They were homemade of cowhide, to which they'd added dried grass for insoles. They called their footgear *porshni*, meaning plungers. At that time, no one had heard about running or hiking shoes in my part of the world, so these were a help. I also added another type of locally available, light and cheap footwear, Asian galoshes that were pointed in front, the favorite of the indigenous people, very useful for fording a stream because you could easily step out of them. During the two weeks of that expedition, I wore out two pairs of such galoshes.

Early on in the 1960s, a group of employees of our Turkmenistan station and Ashgabat Botanical Garden made a joint expedition in the southwestern Kopet Dag, the Upper Sumbar area, to explore the numerous gorges in the valley on the river's right bank. The left side along the river was the border with Iran.

We started out in a truck in the Sumbar River Valley, stopping at a gorge, putting on our backpacks, and walking up the gorge. It was already the hot season, but all of us were young and we did not feel too tired. Later I calculated that during the two weeks we were gone, we walked over 125 miles in those mountains. It was rocky and rough, and our shoes did not last long.

Another long journey took us along the Tupolang River Gorge in the Hissar range in Uzbekistan. I walked along the gorge in the company of several foresters who had donkeys loaded with sacks of flour for their mountain settlements. I attached my backpack and walked with them over 28 mountainous miles. A very narrow path hung high up in the mountains over the blue river far below, from where roaring sounds reached us. Nowhere in Central Asia have I ever seen a more beautiful river than the Tupolang we looked down on. The hardest parts were the "overings." Where there was no ground for the path, road workers had knocked in horizontal poles and stakes, put branches and twigs across a horizontal lattice-like layer of wood. That kind of flooring was called "overing" in Central Asia, and they were often quite worn through. On the poorly maintained ones, our donkeys' legs would fall through. I was a greenhorn, a tenderfoot who felt insecure on that swaying twig mass that looked unstable and unreliable. On the road to the first *kishlak*—settlement—there were at least 25 overings.

The second day, I walked by myself to where wild pomegranates were growing. These wild bushes were the purpose of my long journey begun in Garrigala, going to Ashgabat by bus, from Ashgabat to Dushanbe in Tajikistan by air, from Dushanbe to the Sari-Assiya regional center in Uzbekistan by bus again, from Sari-Assiya to Tupolang Gorge by truck. I walked another 28 miles up the gorge. I walked back by myself, too.

The result of that expedition permitted me to conclude that in the Tupolang River Gorge, I had come on wild pomegranate populations and not the abandoned, cultivated plants that had reverted to the wild, as some researchers had formerly stated. Sometimes you arrive at conclusions in science by going a long and hard way to make them.

From my time as an agronomist in Daghestan, I'd learned how to ride a horse. There, gardens spread over an enormous territory of 1,976 acres, and I frequently needed to be at several locations within the same day. A very hot-tempered stallion had been assigned to me. The horse and I made our journeys interesting.

I used my horse-riding skill in the southwestern Kopet Dag as well. At the station we had two horses and one Bactrian camel that was frequently used for plowing. I remember long rides along the Ai-Dere Gorge and the Sumbar River Valley and back to Garrigala in one day while it was still light—about 56 miles. On another trip to the Ai-Dere Gorge, I slept beside my horse on the ground. Later, O. F. Mizghireva, our director, reprimanded me, "You should not go to the mountains alone, much less to sleep there." As for me, I answered with my favorite saying, "He who was destined to be hanged would never drown in water."

I last rode a horse 25 years ago when I accompanied Moscow botanists and border guards to examine the gorges on the left bank of the Chandir River. The gorges were located behind the so called "border control zone," behind the barbed wire fence. Further to the south, the territory of Turkmenistan rose to a mountain ridge bordering Iran. One could get there only with a special permit issued by the commander of the border guard troops in Garrigala.

On another expedition we had only one jeep. We'd crowded nine people into it, including botanists from Moscow and Leningrad as well as employees of Sunt-Khasardag Zapovednik. I was the only one who knew where to find the endemic Lucerne, a species of alfalfa. We had a young and daring man named Durdi driving us through the treacherous mountain passes. Descending from a pass on a long-abandoned road where boulders and rocks had fallen down was an extreme sport. On foot, we made our way through thickets of thorny blackberries to a little site on the western slope of one of the ravines. *Sinskaya lucerne* occupied a part of this site, no more than 100 square meters. It was a classic illustration of so-called "point area," a plant endemic to this spot and nowhere else in the southwestern Kopet Dag or in the entire world. Most probably, the Lucerne was the result of a rather recent mutation that had not had time to expand to a larger area.

I wouldn't call that trip successful. Winter and spring had been quite dry. Tiny Lucerne plants had dried before they formed flowers and fruit. We had to crawl on the ground looking for the previous year's fruit. Lucerne's adaptive strategy was that separate seeds could sprout at various times, not all the seeds at once. Otherwise, it would have become extinct long ago during dry years, or very humid years, and years with relatively cold winters.

I, of course, had my personal object of interest in that gorge—finding pomegranate bushes that grew there even on the talus or scree. Next to such a bush, I saw an enormous wild grapevine. I'd never seen wild grapes growing on scree anywhere else in the southwestern Kopet Dag.

No matter the difficulties, long and hard travel, getting sick on airplanes, having to find a place to sleep at night, results were what mattered most as I traveled these roads.

# CHAPTER 22

# I Consider Myself a Fatalist

**I consider myself a fatalist** and often feel that I'm rather like Pechorin in Lermontov's novel *The Hero of Our Time*. After all, I survived infant and childhood diseases, artillery and bombs that fell on Leningrad—a large fragment from an anti-aircraft shell landed right in front of me. I survived the evacuation of the city under bombardment. After that, I returned alive from South Daghestan despite many threats by the Lezghins who were grave, non-joking folks.

After the war, back in Leningrad, I jumped clumsily from a streetcar, fell down, got my arm stuck beneath it. It stopped before running me over. Another time, riding a streetcar near the Finlandsky Railway Station, the car jumped off the rails and rolled onto the sidewalk. Everybody was safe and sound, including myself.

As a schoolboy, I was walking in the city when a jar thrown from the sixth floor fell right in front of me. I did not have a moment to become frightened. Once, entering a room in the city, a heavy chandelier fell down right behind me. When an automobile ran over my foot, I only got a bleeding toe. These events happened so fast that I did not have time to become frightened. I really kept getting out of danger just in time—or ahead of it.

I remember my first large expedition in Turkmenistan in 1962, the second year of my post-graduate research. During one month we examined the entire south of Turkmenistan, crossed the desert regions and explored the oases. Many times we had to help our driver Yura Tarasovsky put *shalmans*, long wooden logs, in front of the truck to make it through desert sands. We slept on the truck, sometimes on the ground. I remember lying at the foot of a *barkhan*, a steep sand dune, under bright stars. An impressive black beard grew on my face during that first serious trial of desert and heat.

You always had to be alert in the mountains because anything could happen. Once, out looking for dwarf pomegranate bushes in the southwestern Kopet Dag, I saw a tiny pomegranate bush that grew on the edge of a precipice, 160–230 feet above the gorge bottom. To reach that bush, I had to ascend the side of the gorge about 650 feet, then, by pressing my body into the rock, I moved forward, first facing the rock, then with my back to it—everything about it gave me very unpleasant sensations because since childhood I'd had a phobic reaction to heights. This was a chance, my necessity, to overcome that phobia. I reached that accursed bush, took several cuttings from it, and started back. The way back was even harder. Thank god, bad things in life are fast forgotten—one of our wonderful human characteristics.

Once Natasha Burnasheva and I were exploring an unknown gorge in the Sumbar River Valley. (I still call her Natasha though she is over 60 years old. All the women of our station continue calling each other with their informal names, the way they were addressed in their youth, when we talk by phone from various countries: Natasha, Emma, etc.) We drove the truck to the gorge entry and stopped. I jumped down and made quite an unfortunate landing. My foot caught in a hollow. There was sharp pain. I'd sprained my ankle. The work had to be done, so I limped around the gorge all day.

In another period in my life, in another country, I was looking down another precipitous gorge. We were in Daghestan. Our truck was making its way to the high mountains where the *sovkhoz* beehives were located. The road crossed Agul country. Aguls were a mountain people numbering about 600. The gorge was very narrow. At the bottom, 3,300 feet below us, the river was roaring. The truck had to turn—but to turn it had to back up. One wheel was hanging over the precipice. My heart sank, a very unpleasant sensation, real plain fear! We made it, and later we sat in the mountains high above the noisy river at dusk drinking wonderfully fragrant tea with hot *churek* bread and honey in combs. The aroma of the linden-blossom trees reached us from the other side of the gorge.

In the Megri region in Armenia, I had hitchhiked to reach Nuvadi village, famous for its pomegranates. The car I was in drove along the Araks River bordering Iran, hanging over precipices. I had to repress the impulse to open the door and to jump out onto the uphill side of the road before the truck went over.

In 1975, I took part in the first large scale expedition around Turkmenistan searching for ancient pomegranate varieties. Raindrops began to fall. Eghenmurad, my almost constant driver, had some of his friends riding on the truck's platform and overhead on the roof. Suddenly, one of them knocked on the roof of the cabin so unexpectedly that Eghenmurad braked sharply. The truck ran onto the road shoulder and rolled over on its side. No one was hurt, not even shaken up too badly. Eghenmurad, however, began energetically using words that I didn't usually hear in his vocabulary.

On another journey with Eghenmurad, returning from a long trip in western Turkmenistan, we stopped for tea at the home of someone we'd known from Garrigala. Eghenmurad suddenly realized that the brakes had given out. He asked me if we should stay overnight. I shrugged my shoulders. My principle was to never decide the driver's business. My responsibility was the route and the stops. But Eghenmurad felt homesick. He decided that we could travel at low speed toward Garrigala. About eight miles from Garrigala, we had to cross a hilly area with many little ups and downs. Suddenly, in the total darkness, the light of an approaching car almost blinded us both. With no brakes, Eghenmurad was helpless. Instinctively, he turned to the side. We slid off the road and the truck rolled over. I lost my glasses. Eghenmurad was moaning next to me. All the samples I'd collected had fallen out of the truck.

In pitch darkness, I helped Eghenmurad get out. I had to support him. We limped for several kilometers along the road to the booth of the nearest road station where we had some tea, and decided to move on. A car approached in our direction. I helped Eghenmurad into it and asked the car owner to take him home. It would be easier for me to get to our station alone. The sun was coming up when I reached Garrigala. It was a weekend, everybody still sleeping. I woke up the director, reported the accident to him, woke the mechanic, found another driver and returned to the over-turned truck. I found my glasses and I also collected all my scattered bunches of pomegranate cuttings and eventually we reached the station, our treasures and ourselves intact.

On my way back from another long-distance trip, I reached the bus stop 45 miles from Garrigala. Once there I was eager to be home but scheduled buses weren't due for several hours. I hitched in a car that already had three passengers. We didn't get far. Descending a steep hill, the car drove off the road into an embankment. Something hit me on the head. My ill-starred glasses flew off somewhere. A fragment of glass partially scalped me. In the hospital, I was stitched. Tragically, the little girl who had been sitting in the car next to me died.

I told the surgeon who removed my stitches that this was my third car accident. He said I was born lucky, and that nothing bad would happen to me any more. So far, he's been right.

In 1983, during an expedition with Gennady Viktorovich Eremin from the Crimean agricultural station, we stayed overnight in Koine-Kasir, the last village in the upper Sumbar River Valley. We always stayed in the home of Dovlet-Sahat. He was an excellent man, a real Turkmen, kind and hospitable.

The next morning, he joined us and we started in the truck for the mountains. It was a very steep climb, a hard mountain road, with a very difficult descent to the bottom of the gorge of Koine-Kasir-dere. We each began our work unaware of anything else. The sky was gradually getting darker with grey clouds. Then it turned really dark, and finally it poured. We ran up the hillside. The little brook in the gorge bottom turned into a flooding torrent that rushed forward with a horrifying roar, carrying down boulders, trees, anything in its way. We got soaked to the skin. Driving could not even be considered, so we made covers from pieces of plastic as raincoats and decided to walk to the village, having an approximate idea of our direction. In the blackness, pelted by rain, our feet sliding in mud, I had a heavy heart. I was responsible for everyone. We were advancing very slowly, no stars to guide us, not sure anymore of our direction. At last, far below, we saw lights. A village. I took a deep breath as we began descending to the valley. At the first house, we were offered shelter. We removed our dirt-covered clothes and fell asleep at once for the few hours until sunrise. In the morning, warmed and happy, we talked about the night adventures jokingly and ironically.

The truck we'd left behind arrived with the two older fellows, Geldi and Dovlet-Sahat. They'd slept in the cab. Once it was daylight, they drove to us without much trouble. We had to give our old drivers their due and appreciate their skills. They were connoisseurs of the mountain roads, knew all the twists and turns. They also were wise about stopping when they were in a dangerous place. The expedition went on without further serious challenges. We collected a lot of material. My notepad became thicker with my observations and descriptions.

I've learned recently that Dovlet-Sahat is dead, and with him, the Oriental Persimmon seedling I'd given him to test in Koine-Kosir, 3,280 feet above sea level.

As I recorded data, I always kept notes on observations that appeared not to have immediate relevance to the issues of my direct research but might be of use later. For example, notes from that period led to a paper describing the specifics of the biology of polycarpics, tree-like perennials that bear fruit numerous times—unlike monocarpics that bear fruit but once.

On another journey from Garrigala to Kizil-Atrek where the subtropical test site of the Institute of Land Cultivation was located, we followed a desert road 50 miles westward to Gasankuli. There, in the proximity of the Caspian Sea, sandy soils were very salty. A few gardeners, however, removed the surface salty layer and brought in good soil to plant figs and pomegranates.

The houses were peculiar, built on stilts and pilings because the Caspian Sea level oscillated frequently. Its waters retreated far from the settlement one day and came close to it, almost under the houses, on another.

On the road along the sea, several kilometers to the north of Gasankuli, there were mud geysers. Periodically, with a floundering, gasping, squelching sound, creamy black mud would spring forth and make fountains from their short cones.

Nine miles further to the north, in a village of Chikishlar populated by fishermen and their families, I found remains of ancient gardens arranged on a somewhat different principle than the ones in Gasankuli. Close to the sea, there is always high humidity. At night or during the cold season, the air moisture condenses and a layer of fresh water collects in the soil at a certain depth. If used correctly, that water is sufficient all year round. The gardeners of antiquity practiced a method of removing about a meter of the upper layer of salty soil. In those deep, less salinated beds where moisture accumulated, gardeners grew grapes, pomegranate, figs and other fruits.

From Chikishlar several roads forked. You could continue on the road along the sea or you could take the rough, rutted way that would lead you to the middle of the Meshed-Messerian plateau, an enormous tableland that long ago was known as Dahistan, a large and flourishing country with a developed irrigation system and blossoming gardens. During the eleventh to fourteenth centuries, the Mongol conquest turned it into a desert. Mongols conquered by destroying irrigation systems. There are still remnants of over 100 settlements in the area. To this day, the present-day residents of Kizil-Atrek go to the ruins to fetch large rectangular old bricks that they use to construct their houses. The law forbids it, but that doesn't stop them.

In the middle of the Meshed-Messerian plateau there are

ruins of two ancient cities located close to each other. The town of Messerian (or Misrian, a spelling variant) still has the remnants of the city wall. It once surrounded several minarets and ruined houses. It appeared that no archeological digging had been done there in the last several years. In the ruins of one house, there grew a large bushy malacocarpus, a very peculiar plant with small somewhat sour orange-red berries, for which it was sometimes called "Turkmen cranberry." In Meshed, even less remained, only a mosque without even a minaret.

We continued over the plateau to Kumdag, a small town belonging to the petroleum industry. Earthquakes are common in the region and had damaged this place several times in recent years. The next stop was Nebitdag, the petroleum capital of the area, a snow-white city built in the foothills of the Great Balhan Ridge. (At present, under the reign of Saparmurad Turkmenbashi, the name is spelled "Balkan.") The city of Nebitdag has also been renamed—as have the days of the week and months of the year—by the ruler, often after himself or his family.

I visited that waterless territory many times. A station near Nebitdag utilized cultivation methods based on collecting precious water in trenches. Water collected in a good soil layer over the impenetrable underground clay layer. After trenches were dug, various fruit trees including pomegranates were planted, as well as pistachios, melons, watermelons and other annuals. After several years, I saw that the trees and plants fruited. Everything was going wonderfully well—until years of drought when, for three or four years, not a drop of rain fell on the ground.

Long distances took their toll on our drivers and trucks. I remember something that happened in 1976 after we'd finished our work in the Chardjou region, the eastern-most reach of our expeditions that year. Eghenmurad was a man dedicated to his family; on our long journeys, he'd missed them very much. We hit the road heading home at sunrise. It was about 375 miles to Ashgabat over a monotonous desert road. We saw occasional settlements separated by uncultivated tableland, sandy or covered with *takyr,* bare solid dry clay. Eghenmurad was not in very good health. By evening he was tired and asked me to "Talk, sing, but not to be silent," lest he fall asleep. I sang all the way, discovering as I went that I had quite a large repertoire. I did not have a musical ear, however, and sang off pitch. Eghenmurad was patient with me. We reached Ashgabat that night, slept there, and by the evening of the next day we arrived home in Garrigala.

Eghenmurad was a quiet modest man and a sincere believer, a very religious man. No matter how difficult our routes and routines were, he would never miss his five daily Namahz prayers required of every Muslim. His religion did not make him a fanatic, however. His son was drafted and served in the Army in Russia; he married a Russian girl there, so Mullah Eghenmurad had grandchildren with a Russian mother in Garrigala.

I can't talk about expeditions and not describe the weather. In summer we had frequent *garmsili* or Afghan hot winds blowing from the south. In the Kizil-Atrek region, the hot dry *suhovey* wind began after lunch as if it were scheduled.

Nebitdag, the city of petroleum in the desert, was famous for the 'pleasures' of a desert climate. In July, the streets did not cool down at night after a day of being heated by the sun. I'd leave my room several times to pour cold water over myself. Those were pre-air-conditioning times in the country. When I couldn't bear remaining in my room any longer, I took my folding bed outside under the open sky and slept there without cover, in nothing but my trunks, but even that did not make it tolerable.

Turkmenistan has periodic dust storms, hot winds that bring tons of the finest dust from Arabia and the Near East. The storms are hard to describe. A thick dust layer covers the houses, streets, roads, gets inside the houses. Visibility drops to zero. Your face is black with dust, even under your glasses.

Life is so unpredictable! V. L. Vitkovsky, manager of the national VIR Fruit Cultures Department, was on a flight to visit us. The plane departed from Leningrad. At Krasnovodsk, they were warned of a dust storm in Ashgabat. The pilots appeared to underestimate it. On the approach to the airport the aircraft was thrown against high voltage posts. Only eight people survived. Vitkovsky was among the survivors.

In the southwestern Kopet Dag, spring weather is unstable. In March of 1963, Eduard Lomakin and I went to the mountains on foot. We intended to cover both Khasardag and Sunt. By evening, we had reached Hozli—called "walnut gorge." A continuous spring rain began. We had to find a place to sleep. Under the enormous walnut trees we found a half-ruined adobe structure. It lacked a roof. We were quite wet by then. We took the one log that apparently had been a support for the fallen roof and started a fire that smoldered all night, lying as near as we could, turning our bodies to one side, then another, trying to get warmer. As a memento of that trip, we both developed radiculitis, an inflammation of the root of the spinal nerve, and had to treat our lower backs with a snake poison ointment for months. Still, youth and health prevailed.

During trips in the southwestern Kopet Dag, the most pleasant time after a long hard day on foot was arriving at one of Eghenmurad's friends' homes. He had friends everywhere in Turkmenistan. You ate what they gave you, and then you slept. Early the next morning, women cooked breakfast while the old men went to see to their cattle. Eghenmurad would begin his day praying. We would then eat breakfast, say goodbye to the hospitable hosts, and take off. The sun got higher, and with it, the Central Asian heat began. The heat lasted from the end of May to the beginning of October in Turkmenistan.

Once, Eghenmurad and I were in a little village near the Amu-Darya River when night fell. We stayed with an old man who made *nas*, the local chewing mixture of tobacco, lime and other unattractive items. *Nas* is popular among the Turkmen who chew it, then move the chew under their tongues, and later spit out the greenish brown mass. *Nas* users, called *naskeshi*, said their habit was more price-efficient than buying cigarettes. I refused to taste it. In the morning Eghenmurad woke up with a headache. The heavy smell of *nas* and its components penetrated the entire house.

We then drove along a desert road eastward. Eghenmurad was attentively reading road signs, afraid to miss the turn to Serahs. We saw the turn at last and took the road that led to the border with Afghanistan. It was still a desert road, not a single bush on it. Around lunchtime, we saw a shelter where a bus stopped once a day. We sat down on the dusty ground under the shelter, took out the *churek* bread that Eghenmurad's wife made for him in Garrigala, and also the melon that an old man gave us from the melon field he guarded. It was a good lunch.

On our way to the mountainous country of Kughitang on the border of Turkmenistan, Uzbekistan and Afghanistan, as we were crossing the Amu-Darya River, our ferry broke down. We were stuck in the middle of the river the whole day. The sun was burning us up and passengers weren't timid in the words they used about the mechanics. Everybody was hungry. One fellow made a little bonfire on the deck's metal floor and threw several potatoes in. That experience taught me to have some food with me always. The ferry was fixed by evening and we continued on.

When you travel alone without a vehicle, there are distinct problems. Say that you've made your detailed route and in advance determined places to stay and what you're researching. But those are only the outlines. More often than not, you have to adjust and change your route and give up what you planned. This adds the beauty of spontaneity but also the complications to travel. You always have to carry some food as your emergency ration with you, as you never know where and when you will next eat.

More than once I went to the Mardykan settlement located on the Apsheron Peninsula in Azerbaijan. Many years before, the Mantashev family, rich oil-producers, had had their property in that sandy area. Their Oriental-style palace and beautiful planted garden were surrounded with a high stone wall. When Soviet power came to Apsheron, the estate was requisitioned. In the 1920s, the poet Sergey Esenin stayed in the palace for several days. He wrote his famous *Persian Tunes* there, including "Shagahneh, oh you, my Shagahneh."

The estate was turned over to the VIR Agricultural Institutes, which was how the Mardykan Department came into being. When I went there in the 1960–70s, most important to me were the collections of pomegranate and olives, and the knowledge that Academician Nikolai Vavilov used to come to the station. The place was full of memories. They had a branch of the Botanical Garden there and A. D. Strebkova's pre-war pomegranate collections were still growing in 1960–70s.

They put a folding bed for me in a large room of the palace. Sergey Esenin had stayed in that very room! *Arshin Mal Alan*, the famous Soviet movie, was shot there. Several employees who remembered those days told us stories.

In northern Tajikistan, there is a city called Hojent that during the Soviet era was called Leninabad. Hojent was truly an ancient city widely known in the Orient. The fabled character Khoja Nasretdin was believed to frequent the city. There were famous gardens I wanted to visit.

I arrived in the evening. The orchards and vineyards were closed because it was the end of the day, and even worse, it was a Friday, which meant I would not be able to enter until Monday. I was beginning to feel hungry. Nearby I saw a *chay-khona*—a tea house. Inside on a wooden platform sat a group of friendly-looking young guys. They invited me to join in their circle, asking who I was and where I was from. They were waiting for pilaf to finish cooking. A bottle of vodka stood on the table beside a large *kosah* bowl of apricots. Apricots, by the way, were famed in that area. I am not a person who enjoys vodka, but as the saying goes, you do not enter someone's monastery with your own by-laws—in Rome do as the Romans do—and so I had some vodka shots followed by apricots. After we shared the pilaf, the company wished me good night and left. The tea house manager made a bed for me on the wooden platform. At last, I rested.

I had quite an adventure in 1999. I had been invited to participate in an IPALAC conference in Israel, at Ben-Gurion University in Beer Sheva. ("Beer Sheva" means "seven wells.") IPALAC is subsidized by the USA to expose agriculture workers from developing countries to achievements in Israel.

I flew from Ashgabat to Istanbul without any difficulties. In Istanbul, I was told to go to a certain counter to meet with an Israeli Consulate worker for my entry visa. Nobody was at the counter. A polite young man called the Consulate. The Consul himself answered and said that Sabbath was to begin soon, so nobody could take care of my business. I felt an adrenaline rush and my blood pressure rising. In the heat of the moment, I went to another counter where they were boarding the return flight to Ashgabat. I said I'd decided to return home. The young Turkish girls working at that counter whispered for a while to each other, and then brought another girl, also Turkish, who had some command of Russian. That young girl took me aside and asked me to sit down and wait a little. She went to fetch her boyfriend who would take care of me. She couldn't find him and asked me to wait a little longer. Some time later, a young man approached and introduced himself in English. He was a university student. He and his friends worked at the airport to make a little money. I briefly described my circumstances. Instead of flying to Israel, I said, I was forced to wait for the Consulate to open Monday. I had no place to stay and did not know how I'd spend my days until then.

The young man invited me to stay with him. It was getting dark. We took a bus, then we walked and then we climbed to the fourth floor of a house, and at last we entered a rather spacious apartment. My companion introduced me to his roommates. He said I was a professor. They all kissed my hand. We had a supper, and we talked a little. From my limited resources in the English and Turkmen languages, I found phrases to speak with them because there were many similarities between Turkish and Turkmen.

One of the students had a car and so the next day they gave me a tour of Istanbul, showed me their ancient city and drove me to the Bosporus. We sat in a teahouse on the shore admiring the Straits. On Monday we went to the Consulate. On Tuesday I received my visa, and in the evening they saw me to the airport. These young men from Istanbul were wonderful, well-mannered students who made a profoundly positive impression on me and, through them, gave me a good feeling for the people of Turkey.

As for the conference in Israel, I learned many new things and met botanists from other countries. There were excursions that showed us much of the country, institutes, *kibbutzim*, *moshavim*, gardens and plantations. I visited my son and daughter-in-law who had immigrated to Israel only one week before I arrived. They were just beginning to study Hebrew in the *ulpan*. Before very long, I was destined to join them.

When I look back, I always remember expeditions as a time of embarking on new routes, having new meetings with new people, collecting new samples, and coming away with new scientific materials and new ideas.

In between expeditions, you felt your "anxiety gene" as Nikolai Vavilov used to call it, your wanderlust, rising again. You felt like Pushkin's Eugene Onegin, "feeling anxiety, the desire to change places…"

Gregory Moiseyevich and Emma Konstantinovna, Garrigala, 1962

# CHAPTER 23

# Work Days and People at Garrigala

**I relished expeditions, heading out on the road,** but maintaining a pomegranate collection consisted less of adventure than numerous little tasks that sometimes kept me going around the clock throughout the year.

After the winter ended, I conducted daily research to discover how the plants had done. If the winter had been cold, I ascertained what frost injuries there were. From early spring, I began observing the beginning of growth for each variety, such as leafing and budding time, beginning of blossoming, fruiting. I kept filling my notebooks long after the fruiting period, when the leaves yellowed and began falling. It was very important to measure the crop from each plant separately, to count the number and the size of large, medium and small fruits, of whole intact fruits and fruits harmed by disease or pests. I had to carefully select fruits for morphological descriptions, mechanical and biochemical analyses, for taste (fruit and juice separately), storing and preservation.

We selected and sent fruit to various fairs and shows. Many dedicated lab technicians—frequently under-appreciated and low-paid—made this possible. When the enormous volume of work was completed, I still needed to write my annual report on the collections, as well as accounts of my expeditions. There were always research articles that I'd planned to work on the previous winter.

As a scientist, I was expected to be aware of all scientific news, to know numerous publications in the scientific periodicals and research literature that the station received. This work took entire days and evenings.

During the stressful period preparing the annual account, we forgot that such things as days off existed. After I finished annual accounts, I had to prepare plans for the next year. Frequently I was invited to participate in conferences and seminars, which I looked forward to because I valued meeting colleagues and hearing what was happening in the scientific world, in my specialty, or adjacent specialties, and telling them of my work and achievements. For many years, I did not use vacation time. Every spring I had to find hours to work on our family garden patch, to plant vegetables and tend them. Our family needed me there because salaries were low. A year would pass by, then another, and yet another. So went our life at Garrigala. But we didn't complain. I chose this way of life. I made every effort to have it remain just this way.

Twice in my years in Garrigala, I had to relocate the pomegranate collection to a larger area because it was growing so fast with new acquisitions and varieties that we no longer had room. Relocation was a tremendous responsibility. My experience from the years I'd worked in Daghestan helped me. There I'd participated in starting a new 220-acre quince orchard planted on the area of the previous apple garden whose trees were already uprooted and removed. Quince shoots were planted within one day as all preliminary work had been prepared ahead of time.

As I've written, our Garrigala Experimental Station had the largest pomegranate germplasm collection in the world—1,117 accessions. You may wonder why I repeat "germplasm collection" so many times. (In Russian, the term would be "genofund collection.") Why do we need them? As you know, collections vary greatly. Some are quite peculiar. In Russia, for example, there is a museum of mice, and a museum of wolves. On the island of Bali, and in Hong Kong, there are butterfly parks.

In the opinion of Nikolai Vavilov, germplasm collections are created to resolve important problems, among them, the sustainability of populations and their wide utilization in selection. Maintaining them is a very serious business that is neither simple nor cheap. Russia was probably the first country that by the end of the nineteenth century had begun collections of domesticated plants, primarily varieties of field crops. These studies were begun in Czarist times on the initiative of several enthusiastic scientists who could foresee the future and understand the importance of their collections.

During my years at the station, I wrote approximately 150 articles, made presentations at 54 congresses, conferences, seminars, sessions and meetings—and I wasn't able to publish everything I had prepared.

Pomegranate became part of my name. I was captivated by the fruit. People expressed surprise when I wrote articles on other topics besides pomegranates.

I conceived of the idea to create an International Pomegranate Studies Center on the basis of the Garrigala experimental station. Alas, my dream was not realized. The Soviet Union collapsed. A haven for pomegranates and their researchers was not among the priorities of the day.

I never worked alone but always had help from knowledgeable punicologists, local people, my family and helpers at Garrigala. Many years and much distance has come between us, but I still recall these helpful, interesting, intelligent people and will tell you about those I remember most vividly.

I was fortunate that my very first years of work at Garrigala brought me into contact with many scientists who studied pomegranate culture and selection all over the Soviet Union. Sometimes our paths crossed without any design or plan. We'd find ourselves in the same town or research station. In the evenings, from our humble hotel, we would go to dine and sample local wines. We enjoyed our conversations and meals, talking about everything, becoming close as colleagues. In tea rooms on second floors of the restaurants, they served small pots of tea, and brought little pear-shaped glass cups and some broken sugar lumps. We sat sipping and talking most of the night. The atmosphere of the Orient reigned there—when your thoughts were slow and calm.

But we punicologists understood the necessity of accelerating exchanges between the existing Soviet pomegranate collections that had been created before WW II by scientists who had dedicated their lives to them. The elder generation of punicologists realized that their years were limited, and they soon would have to retire. We knew from bitter experience that when the founder was no longer there, collections perished or were torn apart; hybrid seedling funds disappeared, and no one took responsibility. One was lucky to have publications describing one's work.

Still fresh in my memory are my past meetings with several of the outstanding punicologists in the Soviet Union. It gives me pleasure to write of them.

In the 1930s, Nikolai Vavilov invited Alexandra Dionisovna Strebkova to work for the agricultural institute in Azerbaijan. She stayed for her whole life. When I visited her at the Mardyakan Agricultural Department in the 1960s, she was already a tired, elderly lady, but I was fortunate to be able to see her rather frequently, to visit her workplace in Geokchai, and meet her at conferences. She mailed her articles to me and wrote letters answering all my questions in detail.

Strebkova had become a single parent during WW II. Later one of her two sons died in a summer Pioneer Camp. She named a very wonderful pomegranate after her surviving son Boris. This pomegranate produces a large crop and is frost-resistant. Boris became a naval officer and invited his mother to join him in the Ukraine. But how could she abandon her beloved work? Her entire life was dedicated to Azerbaijan and pomegranates. For over 40 years, she remained at her post where she studied and collected local pomegranate varieties from the Caucasus and from Central Asia.

As she was telling me of her meetings with Nikolai Vavilov, the man we both so absolutely admired, telling me about his letters to her, on the wall was a portrait of a very beautiful, very young girl. This was herself during the period she was telling me about.

In 1970, I received her last letter written from a hospital in Baku. For the first time in her life, she experienced cardiac problems. Later, I was told that no doctor was available at the moment she needed assistance, and thus Alexandra Dionisovna Strebkova died.

As well as my feeling of personal heartbreak, it was terribly sad to see how little her work was appreciated by the local Azerbaijan managers of science after the USSR collapsed. Her pomegranate collections in Geokchai were moved to a new auxiliary locality on the Shirvan Plain. A negligent student ploughed over the site where her famous, almost legendary blue fruit seedlings grew. This rare variety had been grown from the seeds collected from the wild population in the southwestern Kopet Dag. Of course, there remained her books, her articles and the varieties she created. We have our memories of this excellent person, the high quality specialist, who dedicated her life to pomegranates.

I'd actually met Strebkova's husband, Petr Alexeyevich Povolochko, before I met her. He had arrived at Garrigala almost at the same time that I did, but he seemed already very old and tired. Decades earlier, in the 1930s, Povolochko had been a young geneticist. Vavilov was quite impressed and invited him to work for the agricultural institutes. Povolochko researched pomegranate cytogenetics, the plant's cells and its chromosomes. He was the first who separated and studied pomegranate polyploids (the plants with multiple sets of chromosomes.) During World War II, Povolochko was taken prisoner by the Germans. Later he was released, but anyone who had been taken prisoner had a terrible time because former prisoners weren't trusted from the very fact of their having been 'abroad,' however involuntarily. Povolochko's career in science was ended, though it had begun so brilliantly. He could work at our station for only a short time; the administration did not like having him there.

In 1964, during my first years at Garrigala, I was sent to Tajikistan. There I met Boris Sergeyevich Rosanov, the most renowned Soviet punicologist. After that, I saw him almost annually in Dushanbe as well as at conferences and meetings. I consider him to be one of my teachers.

B. S. Rosanov had started working for the Soviet agricultural institutions in the beginning of the 1930s, at the onset of the time of total mistrust and social repression across the entire Soviet Union. The times became his trial, too. Falsely accused, he was convicted and sent to labor camps located in the Central Russian territory. There were many intellectuals like Rosanov who were convicted on false accusations. Many of them were people of science. In the camp, Rosanov met Agnia Sergeyevna Lozina-Lozinskaya, who was then young and very beautiful, as Rozanov later told me.

From the labor camp, he'd been exiled to Denau in southern Uzbekistan where he began to work with subtropical fruit cultures at the agricultural experimental station. For many years, he was obligated to appear daily before the local KGB office. Notwithstanding the difficult circumstances, Rosanov carried out profound research. He wrote and defended his dissertation, began his selection work on frost-resistant varieties and then on soft-seeded pomegranate varieties.

Several years later, B. S. Rosanov was permitted to move to neighboring Tajikistan where he became a professor and was awarded the honorary title of Meritorious Scientist. The scope and productivity of his work grew. He published his monograph on pomegranate culture in the USSR, summarizing his own research on the biology and the selection of the fruit with fourth-generation seedlings. His work resulted in soft-seeded as well as semi-dwarf varieties that could be better protected during the harsh winters in Tajikistan. Though summers are hot, much of Central Asia has severe frosts in winter. Pomegranates inevitably get frost-bitten or freeze completely. Without winter covering, pomegranate culture becomes absolutely unprofitable.

Rosanov's new and promising work on soft-seeded varieties slowed in 1972 when he became ill. After prostate surgery, local surgeons refused to use a new effective medication made in France that Rosanov's daughter had brought from Moscow. The life of Boris Sergeyevich Rosanov ended at the height of his creative potential and energy.

One of his best seedlings in Garrigala had a magnificent dark red juice, almost blood-red in color. His disciples gave the variety the name, "In Memory Of Rosanov."

I had actually met Agnia Sergeyevna Lozina-Lozinskaya in 1945, when I was a school kid in Leningrad, passionate about plants. By then, she was already a senior researcher at the Botanical Institute of the USSR Academy of Sciences, and the well-known author of many works on plant systematics. For several years, she followed my progress and my life. She was a person with a big heart. She understood my monetary hardships and once suggested that I dig a test site for her. I could hardly cope with the work, but she paid me. Years later, I realized she'd paid from her own pocket. It was my very first earned money, 30 rubles, no less, no more!

I finished studying at the Leningrad Vocational School of Green Construction that trained medium-level specialists to work on urban planting and landscaping. I graduated *cum laude*—such good grades should have given me the right to be admitted to any university or institution without taking entrance examinations. But I was not admitted to the Leningrad University Biology Department. There was an unwritten directive to admit no Jews for a number of professions at colleges. The next year, Agnia Sergeyevna offered me assistance and protection to apply to my desired biology department, but I refused. For many years, I felt hurt that I'd been rejected the first time.

By the end of the 1980s, Tajikistan was being fought over by local nationalists who destroyed the experimental stations. Fortunately, we already had Rosanov's best selections included in Garrigala's pomegranate collection. I began sending them to many research institutions in the former Soviet Union as well as to foreign institutions.

In Armenia, Emma Airapetovna Gabrielyan-Beketovskaya worked on pomegranates for many years. She collected and researched the local varieties and always assisted me in my work when I visited Armenia. I am sad to say that one day I arrived at her Institute and learned that she was no more.

Other centers of agricultural research were better set up than we were at Garrigala. When researchers from the Nikitsky Botanical Gardens in the Crimea visited us, they noticed our spartan conditions and work arrangements. It's true that among Garrigala employees, we had 'natural selection.' Only the fittest survived, the ones who were really enthusiastic.

As I traveled all over the Soviet Union, from Azerbaijan to the Caucasus and the Ukraine, then returned to Garrigala for all the work the station demanded, I couldn't have accomplished my goal of collecting pomegranates "from around the globe," as Nikolai Vavilov used to say, without the assistance of co-workers, family, and often visitors who came to Garrigala and became our volunteers, part of our team.

When I first came to the station in 1961 and was just becoming familiar with the collections, I saw a beautiful young Turkmen woman working on the pomegranate site. Her name was Tajigul Murodova. She had a dish of a cream-like mixture of dirt and pesticide in her hand with which she covered the calyx of each pomegranate fruit. The idea was that such an operation would reduce the harm done by *Euzophera punicaella*, a pomegranate pest. I introduced myself and we talked. Tadjigul happened to be my age. I kept up on her life, her family and her children. Years later, her husband was killed in a car accident. She was already a grandmother when this happened. Tajigul aged, of course, but her face remained beautiful with its noble intelligence and wisdom. After she retired, she rarely came to the station. She said her legs hurt, but more than that, she said she felt like vomiting to see how the station, to which she'd dedicated her life, was neglected.

Maria Antonovna Nastkalich.was a lab assistant when I came to Garrigala. She worked there until her death, caring for the collections in the field. She had to work to support her family and stayed until she was 70.

My wife Emma was my primary helper. Like other wives at

our station, she took upon herself the entire load of raising the children and doing the housework. At the same time, practically all the time, she assisted me in my work. She was 'in-house' editor for my articles. When I couldn't see errors, she could. Emma typed my works at home on our quite old typewriter. We never had computers at our station.

My son, Alexander, called Sasha, graduated from a very good high school in Ashgabat and decided to become a physicist, but when he applied to the institute of his choice in Leningrad, a secretary marked his application packet so that he was denied admission. He began working in my lab as an assistant at the beginning of pomegranate harvesting, recording everything, having a huge responsibility. A short time later, on the eve of his birthday, Sasha received a summons to appear before a military draft committee. He was drafted to serve in the Soviet Army, but fate was kind and did not send him to Afghanistan. One-third of his buddies who went returned in coffins. After military service, Sasha was admitted to the Leningrad Polytechnical Institute and became a radio-physicist. He lives with his family in Israel where Emma and I have joined them.

My daughter Elena finished high school in Garrigala and decided to follow in my steps. She applied to the Leningrad Agricultural Institute, where, as it had happened to me, she was refused admission. Like her brother, she began working for my lab until a year later, she entered the Moscow Temiryazed Agricultural Academy. Presently she lives in Moscow with her husband and their 16 year-old daughter. My granddaughter Katya has toured Europe during her autumn vacation, and with her parents, has visited us in Israel. It's a different time. It's much easier to see the world now.

We always had visitors to our station who helped with our pomegranate collection. B. V. Belyakov, a Moscow journalist from *Pravda* as well as a professional historian, was about to be sent to Egypt. The government wanted him to have a preliminary idea of Egyptian landscapes, so they sent him to Turkmenistan, as if it were like Egypt! He liked everything about our work, so I asked him to collect Egyptian pomegranate seeds for us. From those he sent, we grew seedlings that we planted in the collection. I was still in Garrigala when their first fruit ripened.

Professor P. V. Florensky, a geologist, visited the station. He gave us a lecture and told us the news from the outside world. We were in the backwoods, in the back of beyond, and relished hearing about scientific events. Professor Florensky had a job in Algeria. Naturally, I asked him to send us Algerian pomegranate seeds, which he did, and which I added to our collection.

Oleg Yurievich Orlov, a zoologist and professor at Moscow University, frequently brought his students for practical work in our area. At Orlov's request, Boris Nikolayevich Veprintzev, a specialist in bird calls, gave me pomegranate seeds from the Himalayas that had many similarities with a widely distributed Kazakh variety.

Alexander Yurievich Kulenkamp, a fruit specialist and professor in Moscow, was a frequent visitor at our station. He asked his foreign students to send pomegranate seeds to me. One day a parcel arrived, a plastic bag of viscous thick liquid with pomegranate seeds in it. I tasted the liquid—it was honey! A Sri Lankan student decided that packing seeds in honey would get them through customs, but needless to say, it killed them.

I never was given a chance to take part in expeditions to foreign countries, but I learned from visitors and my own reading about pomegranates outside our boundaries.

Afghanistan interested me greatly. Vavilov had visited that amazing country and written that Kandahar was renowned for its pomegranates and its pharmacies. He wrote that the pomegranates he saw were the size of a child's head. He also described the celebrated seedless pomegranates that grew in the royal gardens.

The first time that I tasted seedless pomegranates, I had the impression of eating butter-soft fruit. Probably such pomegranate mutations have occurred more than once.

Over the years, Soviet researchers brought back Afghan varieties that we added to our Garrigala collection. One variety, Afghansky, had fruit of medium size that ripened early, late August to early September, and had soft seeds and light pink arils.

Even in Vavilov's time, traveling in Afghanistan was dangerous. That troubled country with its botanical riches is presently beyond the reach of researchers. Let us remain optimistic that one day we will be able to travel there, and that the next generation of punicologists will complete the detailed research of Afghanistan pomegranates with all their genetic riches.

I have only sketchy ideas of pomegranates in the United States. I know they are cultivated mostly in California, and that Paramount Farms manages an industrial-size crop, where orchards yield an average crop of 10–12 tons per acre. Their scientists have been testing the newest methods at the Volcani Center in Israel, covering pomegranate plants with nets of various colors.

Pomegranates are a leading fruit culture in Israel. Each year, 4,000 tons of pomegranates are sold, 50% at Rosh Hashana for the Jewish New Year. In the Volcani Center experiments, many plants of various kinds (for example, myrtle) have responded quite differently depending on the covers used on them. Plant size changed, became larger or smaller, as did the quality of the crops. The reduced evaporation saved 20–25% of the water needed for irrigation. Covering may also reduce the number of sunburned fruits, improve fruit quality—fruit color, arils and juice—and increase productivity.

I first visited Israel in 1999 to participate in a conference. At that time, I promised to send 40 of our best pomegranate varieties and 40 olive varieties to the Ben-Gurion Negev University in Beer-Sheva. Upon my return to Garrigala, I did just that.

In 2003, I visited the site in the Negev Desert where Dr. Wiesman had planted my 40 best soft-seeded pomegranate varieties developed at our Turkmenistan station. The pomegranates were being irrigated with a drip system. The three year-old trees already produced fruit. It was an intense production process because of the high interest in the antioxidants in the juice.

But the story of Garrigala's pomegranates in Israel isn't all a happy one for me. I discovered to my surprise and dismay that some of the best pomegranates I'd sent from Turkmenistan, the varieties that were growing and producing desirable fruit, were now identi-fied by numbers rather than the names I'd given them. By refusing to reveal the names of the varieties that I had identified them with, they became inaccessible to researchers wishing to propagate them in California. I'd never experienced such a kind of possessiveness motivated by greed. My life hadn't prepared me for this situation, but I decided that it takes all kinds to make a world. I couldn't have done things any way but the way I did.

After Turkmenistan expedition, 1962

# CHAPTER 24

# Dark Clouds over the Soviet Border

**It so happened that when wild pomegranate populations** I wanted to find grew along the southern borders of the Soviet Union, we had to request permits from officers of various ranks among the border guards. Permits occasionally would be issued promptly, and sometimes not. A good thing was establishing relationships of trust, almost friendship, with the officers along the borders of Turkmenistan. "Expedition brethren" we called ourselves. Local people treated us in a friendly way, served us meals and provided assistance. We, in turn, brought some variety into their monotonous routine.

A fellow Leningrader commanded an outpost in the Chandir River valley in the southwestern Kopet Dag. When we arrived, the Lieutenant was in a melancholy mood. His wife had recently left him. The Turkmen hadn't provided conveniences for her and border guard society hadn't been to the city lady's tastes. We felt embarrassed to burden this sad man and decided to sleep in our bags near our jeep. It wasn't to be a quiet night. A border invasion set off an alarm and the soldiers rushed out. It might have been a porcupine who touched off the electric signal going under the wire, or a leopard jumping over. It wasn't likely that any person was trying to cross illegally there.

After being awakened several times, we accepted the Lieutenant's invitation to sleep at the outpost on regulation iron cots. We ate supper together and talked late into the night. He mentioned that until very recently pine trees had been growing on the Turkmen side of the ridge. The Iranians had crossed over, cut down the pines, and carried them away. That got our attention. Pine trees did not belong to the flora of Turkmenistan, nor to Central Asia. But we weren't permitted to get that close to the border. We botanists asked a soldier to bring cones from the spot the pines had grown—by examining them, we'd identify the variety.

We worked at that outpost for several days, exploring the gorges along the Palizan Ridge, going over the control line. I was searching for wild pomegranates while my companions were interested in other local flora for herbaria. My wild pomegranate samples from that trip have been included in material that I used for a number of papers. But the pine cones never reached us, and we weren't able to include the pine variety in Turkmenistan's list of flora. No botanical sensation took place. Later, I learned that the Lieutenant had made some professional mistake and been transferred elsewhere.

We did not have any problems with the border guards in Tajikistan, either. I made a trip to the Dashtijum *kishlak* in the Kulab region, an area that later became a national park and included wild pomegranate populations. From Dashtijum, we drove along the border between Tajikistan and Afghanistan—just on the other side of a narrow river. I'd read about that natural boundary in accounts by punicologists before WW II. By the time I arrived, the settlements had been abandoned, but pomegranates were thriving in the deserted gardens. Conditions were quite favorable—appropriate microclimate and thick layers of rich soils deposited by the river.

Another Tajikistan trip took me to the southwestern Pamir. From Kalai-Humb *kishlak*, the road went along the Pyanje River bordering Afghanistan. The many little gorges descending to the river had small populations of wild pomegranates as well as a significant variety of cultivated pomegranates.

I didn't have problems when I traveled along the Soviet border with Iran in Armenia—at that time a Soviet republic. The Megri region was very interesting to explore. The road from Megri to Nuvadi *kishlak* was along the bank of the Araks River, itself a natural border between Armenia and Iran. Some parts of the road hung over the river. Nuvadi, an Azerbaijani enclave among Armenian settlements, turned out to be a real museum of pomegranate varieties. Later when the Soviet Union collapsed, the Azerbaijani residents of Nuvadi had to walk from their settlement to Azerbaijan carrying their children in their arms. The conflict with Armenia dispossessed these people forever.

I was with a colleague driving to Kizil-Atrek along another border with Iran in the direction of the Caspian Sea. What interested me were pomegranate varieties growing in the few settled places along our route. On this trip, something went wrong with our jeep. We could see a border outpost in the distance, so I walked to it to ask for help. The guards invited us to join them for dinner and sent help to our car.

The situation around the borders in Azerbaijan were different. I was interested in the Talish area adjacent to Iran. To get there I took a bus from the town of Lenkorani that went toward the south. There I was stopped at a forward border post in Mollahkend and was told that I needed a permit from the Lenkorani border guard, even though my status as a legal resident of Turkmenistan gave me right of access to all the frontier/border zones of the country. They didn't care about that, so I returned to the Lenkorani border guards and requested a permit to go to the Talish area, the town of Astaru located at the border. The conversation reminded me of an interrogation. The outcome was negative. They denied me a permit for the first time in my life.

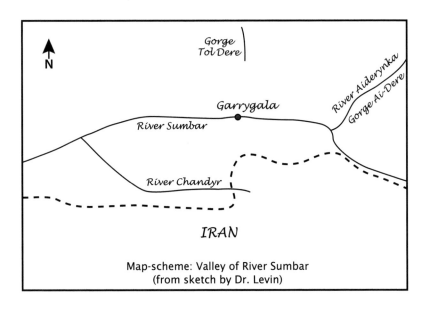

Map-scheme: Valley of River Sumbar
(from sketch by Dr. Levin)

CHAPTER 25

# Two Astonishing Gorges

**"Along the Valleys, Along the Mountain Foothills"** is the first line of a song about the 1918 Civil War in the Amur River and Vladivostok Primorye area that always comes to me when I think of trips into mountain gorges. I made many of them to see how the wild pomegranates in a specific area had coped with the severe cold of previous winters, when temperatures dipped below their level of frost resistance. I wanted to assess damages in different gorges.

Severe winters were rare in the southwestern Kopet Dag, happening only two or three times a century. In the winter of 1928–29, temperatures dropped to record lows. During a second catastrophic winter, 1968–69, cold air from the mountains descended to the valleys of the Sumbar River and the Chandir River, and pomegranate orchards there suffered very badly, sustaining much damage. In the majority of cases, pomegranate bushes had to be sawed off at their lowest part. We then waited for the new shoots to grow to keep the right number—the optimum was four to six trunks—for best size and crop. The entire process of restoring these pomegranates took four to five years.

I patiently waited for the weather to become warm so that it would be clear what amount of damage the wild pomegranates in the mountains had sustained. One year, Vasily Andrianovich Nosulchak, who headed our grape laboratory, went with me to the

155

gorges in the Chandir River Valley. I was checking and noting the frost-damaged pomegranate plants, while Nosulchak observed the wild grapes so numerous in the southwestern Kopet Dag. We were working steadily every day, examining one gorge after another. I chafed at the time it took because work at the station needed me there as nearby, my companion worked patiently on his own botanical interests.

Before that expedition, I'd felt that Vasily Andrianovich was a reserved person. I found out that he was a man who concentrated on his work. In the gorges, at the hottest time of the day, we would find a shady spot and have a picnic. He had simple meals: a can of beef with bread. He washed his can and boiled his tea in it. He told about himself, his childhood in the Ukraine, his military service in the Army, his college in Odessa and his postgraduate school, the books he loved and everything else.

In later years, Nosulchak became our station director and remained head of the grape laboratory. I remember the long winter evenings when we were working on our annual reports. From our labs next to each other, I heard him working and whistling. I was doing my work humming. We didn't distract each other from our annual accounts.

Nosulchak's expeditions and acquisitions from other research institutions in the USSR and abroad increased our station's grape collection to 1,000 varieties.

After the Soviet Union collapsed, the agricultural institutes lost all their grape collections in all of the stations. Nosulchak, after 30 years of work, had to leave Turkmenistan and transfer to the Crimean station in Krasnodar where he would have to create a new collection of grape varieties. There was a grandiose plan to develop a collection of 5,000 varieties within ten years—that meant 500 varieties a year on average. I do not know whether Nosulchak kept up the planned tempo.

After I moved to Israel, I wrote to Nosulchak and offered my help developing his grape collection. Nosulchak sent a list of varieties he especially wanted. I passed on his list but I don't know whether anything came of it. Nosulchak was always a very hard working person, always overburdened with work. And at his new location, there wasn't enough money to support him or any staff. All researchers had to get a second or third job just to survive. I heard that people at the Crimean station grew vegetable crops and sold seeds to survive—such was the new economic reality.

I made annual pre-New Year trips to Bagandar Gorge not far from Garrigala. The entire *kishlak* enjoyed pure water from its mighty spring. We barred entry to the gorge to prevent cattle from reaching the spring and polluting the drinking water. As a bonus for us, the plants in that gorge remained intact, growing in practically national park conditions. During my excursions on New Year's Eve, I made notes of plants already blossoming in December, or still having their late autumn blooms.

There was a rather narrow, creviced gorge that took you to the east of the Bagandar spring. Thick blackberry bushes completely blocked my entry. To get into that crevice, I had a difficult climb along its steep walls. A lot of wild sorrel grew on the bottom. And at its far end, a large wild pomegranate bush stood by itself. Even at the New Year it still had rather large fruit, uncracked rinds and a very good sour-sweet taste. I had taken cuttings from that bush many times. Unfortunately, they did not grow well in our nursery. It was one of those Donguz-Nar pomegranates, according to the folk Turkmen classification, meaning *Pig Pomegranate.*

Further still into the crevice on its steep ascending wall and in between its cracked rocks, there grew several pomegranate bushes with very bright-colored, small fruit. They were a natural decorative form that were easily adapted to gardens and public parks in regions where winters weren't too cold.

At that time of year, you could always find Michelsens' crocus blooming, tiny graceful plants. To the best of my knowledge, it was their westernmost location in the Kopet Dag. There was another crocus with yellow flowers in Turkmenistan that was distributed over the huge space of Badhiz in southwestern Turkmenistan. I visited Badhiz to learn more about a pistachio savannah as well. The region made an unforgettable impression on me.

Unfortunately, when the Soviet Union collapsed, our regular trips to the gorges were stopped by our newly appointed director. He regarded transportation expenses as costing too much for his plan. A great pity. The longer the observations continued, the more valuable conclusions could have been drawn.

There's so much to write about the mountain gorges of Turkmenistan that I shall take my reader to a few of them in more detail. Let me say that I am obsessively attracted to mountains. Even now in Israel, I always turn on TV when Alexander Gorodnitsky and other bards sing about mountains. Those songs are in my soul—they've been so since my youth and will be with me forever. My expeditions permitted me to explore many great mountains in the Caucasus, Carpathians, Kopet Dag, Kughitang, Hissar, western Pamir and western Tian Shan.

The mountains were my life for over half a century. I still make comparisons. In Israel, the mountains of Sinai remind me of the Kopet Dag, and so do the mountains of Judea. The northern Galilee, Har Karmel and Golan Heights also evoke reminiscences, especially in spring, when they look like the Caucasus or the Carpathians. I can say with Robert Burns, the great Scots poet, "My heart is in the highlands." With sadness, I think about my youth when I had energy and strength, and when life gave me the opportunity to breathe the mountain air and seek the endless vistas. Those were times when I felt that everything was still before me, that life was still to be experienced. Nonetheless, this isn't yet the end. As Vladimir Visotsky sings: "Better than the mountains you have visited only are the mountains that you have not yet gone to."

**Yol-Dere Gorge.** I was once working in the Herbarium and Flora Department at Garrigala, studying pomegranate samples we'd received from Vienna and London. Work was going well, everything was interesting, when suddenly, from among the samples from all over the globe, I came across collections dated 1912 by Alexander Ivanovich Mikhelson from the foothills of Sunt Mountain and the Yol-Dere Gorge located less than ten miles from Garrigala!

On my first visit to Yol-Dere in early May 1961, the gorge was to my eyes as impressive as it had been described by those who had exclaimed over its beauty. I stood under a bright blue sky, with the murmur of water in its brook below, the slopes covered with uncountable varieties of beautiful blooming plants.

Where I entered, an enormous rock fragment most likely weighing hundreds of tons had fallen. The entire area is prone to earthquakes; large quakes occur several times a century while smaller tremors happen every year. The strongest earthquake recorded occurred in 1948 near Ashgabat, the capital of Turkmenistan, killing more than 80,000 people. My future wife Emma, a girl at that time, was injured, as were her mother and father. The earthquake killed her sister and her aunt.

That first time in 1961, I was with T. N. Ulyanova, a postgraduate. We set our tent near the brook. For several days we collected annual bean plants. Everything in the entire country was new for me. I greedily absorbed the surroundings. Not being accustomed to night sounds, I thought that over the murmur of the brook I was hearing an animal moving about. I concluded by our dog's behavior and by paw prints we found that a leopard had come quite near. I did not have a chance (or misfortune) to see our visitor face to face. When finally I fell asleep later that night, I was awakened by a bat landing on my face. The sensation was not a pleasant one.

Since that initial trip, I would come to the gorge at least once a year. For many years it was a dream place for many visiting botanists. New plants were frequently discovered. Yol-Dere Gorge was thought to be the richest for plant varieties among all the gorges in southwestern Kopet Dag.

Later, Galina Mikhailovna Proskuryakova, a TV anchor of *The World of Plants* came to our station with her crew and filmed our pomegranate collection. My last meeting with Proskuryakova was in 1987 in Moscow at the Fifth Congress of Geneticists and

Horticulturalists, convened to commemorate the centenary of Academician Nikolai Vavilov's birth. She died one year later.

As always, everything has its beginning and its end. My last trip to Yol-Dere Gorge was in early May of 1998. Professor Bill Feldman, visiting from the USA, accompanied me. The Yol-Dere brook was full of water. The gorge was beautiful, the way it always was in spring. Tulips and crocus, however, were already almost gone; we could hardly find one to photograph. There were other changes, too. I didn't find the wild peas that had been there before. Nor was the Oval Listeria solitary orchid, that rarest variety in the southwestern Kopet Dag, where it usually appeared. The waterfall in the gorge was tiny, while in previous years it had been quite full. So many changes as years passed!

Bill Feldman sent me his article describing that trip with photos published in *Desert Plants* magazine. He had truly liked our gorge.

Pomegranates grew in Yol-Dere but not abundantly. The bushes grew along the brook, on the slopes and at the very entry to the gorge on the southern slope where the plants were rather short and stocky. Many of them formed groups of low-growing bushes attached to each other underground. Pomegranates weren't thriving on the slopes because they needed more water. The bushes in the gorge bottom along the brook were tall, but even there, they had only scattered and weak fruit. I felt that low levels of light in the gorge bottom was determining the pomegranates' poor growth.

**Ai-Dere Gorge** had been visited at least a century earlier and was already known for its beauty and climate. In 1891, a secretary to a Russian military man wrote,

*Ai-Dere gorge known as 'Bear Gorge' was covered with woods for about 14 versts. A mass of almond trees grew in abundance along the brook and on both slopes. All almond trees were heavy with fruit that were predominantly bitter. The residents of the nearby Nuhur settlement picked only sweet almonds. Annually up to 200 – 300 puds of regular almonds could be harvested, and 1,000 – 15,000 puds of bitter almond.*

The secretary then noted that the climate and landscape made cultivating the gorge possible. At that time a "pud" was an outdated Russian weight measure that equaled approximately 35 pounds. A "verst" equaled about two-thirds of a mile.

The following year, reports of jujubes (*Zizyphus jajuba*, also called Unabi dates) and walnuts were added to our list of attractive native fruit trees. All horse-cart makers from the Nuhur *kishlak* and Ashgabat took their wood from there.

From the 1930s on, researchers at our Garrigala station studied the gorge, particularly focusing on useful plants. The almonds, for example, that we brought to our station and planted became the heart of our collection. One of the earlier researchers, Petr Nikolayevich Bogushevsky, had selected about 400 promising different almond varieties and planted seedlings that made our almond collection at Garrigala so rare.

Bogushevsky and his wife starved to death in besieged Leningrad during the war. I'm sorry to write that the destiny of Bogushevsky's almond collection at our station was also tragic. Much of it was cut down by the local population for fuel during the war. After the war ended, the collection was begun again.

Anatoly Valerianovich Gursky explored the walnut thickets in Ai-Dere Gorge where he found many very good varieties of large fruit. Gursky went on to create a botanical garden in the Pamirs where he spent 25 years. When he died, the local people, who had called him Abdullo, mourned his passing.

I traveled the entire length of the gorge on foot or horseback. The numerous wild kin of many cultivated plants continued to interest me. I remember these trips well.

In summer, the narrow Ai-Derenka rivulet was shallow and in places you could leap over it from one bank to the other. In early spring, however, Ai-Derenka swelled to a mighty stream. There were 33 fords to make across it. You had to walk on the terraces—quite arduous and quite slow. My companion, N. M. Minakov, once fell in and had to wring out his clothes in the rain. I tried getting a photograph but failed.

On field trips with respected professors and young students, we spent long evenings around campfires sharing stories. Many of them turned around the tragic battles of the past against Lysenko and his followers. Lysenko's doctrines and Stalin's support of them caused many deaths and set back biological sciences. The 'ruling' scientists insisted on their medieval views as long as they could. This campaign caused such damage!

Usually I would get to Lower Ai-Dere *kishlak* by evening time, stay the night at the house of Muhammed or his son—Muhammed headed a department of a subtropical *sovkhoz*. Both were very hospitable, a quality shared by all Turkmen, and also by all the Muslims in whose houses I stayed in Central Asia. In the morning, while it was still cool, I began my walk up the gorge. I came first to groves of walnuts managed by the local forestry department. Further along the path, thickets of blackberries, hawthorn and wild grapes grew densely. Among blackberry bushes you could see a group of plants that were less fruitful hybrids of the local blood-red blackberry and the gray, dove-colored blackberry, typical of higher locations in the southwestern Kopet Dag. Soon you began to see wild almond trees on the slopes. By then, the day was growing hotter, time for a picnic under a giant plane tree near Miraji, the abandoned settlement located in the middle gorge. The Ai-Derenka brook murmured, the air was sultry and torrid. The water that I put in the middle of the bonfire in the *tuncha* (a peculiar local teapot shaped like a jar and made of metal) began to boil. We ate a modest lunch and went on our way.

Occasionally we had unpleasant encounters. Once I forded the brook and got to the other side while Minakov was leading the horse across the water. Suddenly a cobra threw itself from the bank into the water and glided between my companion's legs. You had to be ready for snakes because they were everywhere.

Further up, there were more plane trees, as well as jujube groves and the remnants of abandoned gardens, traces of past settlements. There were many such settlements or *kishlaks* in the Sumbar River Valley and along its tributaries. Wherever water was, Turkmen would have made a home for themselves and their clan.

In the 1930s, the Soviet Union controlled Central Asia and collectivized agriculture in Turkmenistan. Many Turkmen could not accept losing their traditional way of life, their cattle or their land. One night an entire settlement would flee with all movable possessions, and cattle, to cross the Sumbar River to neighboring Iran. Descendents of these Turkmen still live along the entire border. After the Soviet Union collapsed, they were permitted to visit family on the Turkmenistan side, and vice versa.

Further up the Ai-Dere Gorge, you found more and more walnut groves. Sapar-Bahar, a large abandoned *kishlak*, the last in the Upper Ai-Dere, was a good place to stop for the night. There were many blackberry bushes along the brook. One had the largest berries I ever saw anywhere. You came across giant walnut trees, each with its own name: Kara-Koz (Black Nut), Shah-Koz (Royal Nut), Kovi-Koz (Best Nut), Kolin-Koz (Thin-Shell Nut). The local population knew each one of the trees by name. They were several hundred year-old trees and still very fruitful.

In the Dayinder-Dere lateral gorge, there was an abandoned garden where varieties of local plums, pears and almonds grew. In my first wanderings in the southwestern Kopet Dag, I also found very thick *kerkay* trees, a Turkmen maple. It was said that no wood burned hotter—sadly, that quality doomed these beautiful giants to be used as fuel.

As we continued our way upwards along the Ai-Dere Gorge, we came to a fork where two gorges merged, Kalin-Koz and Kara-Su. Kalin-Koz was especially beautiful with shadowy nooks in semi-darkness, tufts of hanging moss on the tree trunks. The name given the gorge came from a tree named Kalin-Koz, famous for its *kappa*, or gummy excrescence, much valued for expensive veneer in finishing furniture. Before the 1917 Revolution, Central Asian *kappas* were exported to France in great quantities.

Walnuts were the main product of Ai-Dere Gorge. Thirty years ago, the gorge was declared a part of Sunt-Khasardag National Park, but that hasn't prevented poachers from taking off with walnuts. Walnuts are also the main nutrition for many mammals in the gorge, primarily hogs and porcupines. Natural propagation of walnuts has become compromised, partly from grazing and the *sels*, or mountain torrents, that rush through the gorge damaging the trees. The Kopet Dag bear is extinct now. Leopards have become very rare. Poachers kill the leopards and there's no food for them. Ranchers don't appreciate the leopards' hunting cattle for food.

The climate in the Ai-Dere Gorge was quite mild, not too hot in summer, with cool nights thanks to the high elevation. Here was where the best pomegranates in the entire Sumbar River Valley were cultivated. The area was beautiful, a real pearl.

My specialty, wild pomegranates, were growing but not spectacularly, not in the way they grew in gorges the Turkmen people called *Narli* (*narli* means pomegranate) in the Sumbar and Chandir River Valleys. In the Ai-Dere Gorge, pomegranates grew near the lower Ai-Dere settlement. Though the varieties there didn't especially interest me, every year I surveyed these gorges to see how the plants had done after the winter. I concluded, as a result of many years monitoring, that the climate wasn't ideal for the wild pomegranate population.

In the Sapar-Bahar abandoned *kishlak* in the upper Ai-Dere Gorge, many gardens with local pomegranate varieties kept growing. I thought that the Nuhurians had brought them in the past from the nearby Iranian areas. Among the local varieties, there were occasionally excellent ones. The majority had been domesticated from the wild. I came across one excellent variety that I'd also found at the Koine-Kasir settlement, 3,280 feet above sea level in the upper Sumbar River region.

The coincidence could be explained by the ethno-cultural similarity of the Nuhurians residing in rather isolated mountain regions of the Upper Sumbar and the adjacent territories. Scholars call that area Nuhuria. The Nuhurians always lived in the mountains. Their isolated existence preserved their peculiar ways of living and culture to the present day without changes. They have specific physical characteristics as well as linguistic features. They resemble Semitic people. Their women are very beautiful. Some ethnographers believe that the Nuhurians preserved the unchanged purity of the Turkmen group itself. Others disagree. There are some Turkmen tribes who consider Nuhurians to be non-Turkmen, or not real Turkmen.

Ai-Dere Gorge represented the eastern limit of commercial pomegranate cultivation in the southwest Kopet Dag region. Above that gorge in the Sumbar River Valley, pomegranate plants suffered from cold winter temperatures.

CHAPTER 26

# The Rock-apple Tree and Natasha; to Kughitang

**At Garrigala, we frequently shared expeditions** to explore the southwestern Kopet Dag. Vehicles were limited so we went together and then made our separate ways depending on our interests.

This time, we started as a group in the Ghyaz-Dag Range that separated the Sumbar River Valley from the Karri-Nuhur Plateau. According to the legend, it was on this plateau that Alexander the Great of Macedonia fought against the Parthenians. Naturally I was looking for pomegranate plants that I monitored annually. Natasha Burnasheva, our specialist on seed cultures—apple, pear and quince—wanted cuttings from the unusual "Rock-apple Tree" that grew in that area. Eduard Lomakin, our young manager from the fruit culture lab, and Geldi Tajiev, our driver, made up the party.

From Koine-Kosir settlement the road ran about 3,300 feet above sea level. As we proceeded to the Almali (Appled) Mountain Massif, it became more picturesque. Archa Turkmenian juniper forests covered the mountain slopes—these junipers were becoming rare in the world, and even here we saw stumps and other signs of illegal cutting. The local people used the wood for the ceilings in their houses.

Almali Mountain and the nearby gorges were covered with numerous varieties of trees and briars. Apple trees, spread on rocky outcrops, didn't grow tall. They were almost unnoticeable from a distance, so it was hard to recognize the place where we'd found the special one in the past, that bizarre apple tree we called the "Rock-apple Tree." We had one sample of it hanging on the wall in the station lab. This trip we wanted more for the station. But first we had to find it.

The weather wasn't helping. In the Sumbar River Valley the sun had been shining brightly, but now we'd reached 6,500 feet and rain fell. For two days it rained. We put our tents over the truck's bed, got into our sleeping bags and slept. In the middle of the night, rainwater collapsed our shelters and we were flooded. Everybody but me had worn-out raggedy sleeping bags. They were almost floating in water. Geldi, the driver, was also unhappy in his tiny truck cabin. The next morning he said he was very cold.

At sunrise, the rain let up and we went to gather brushwood for a bonfire. I picked up one juniper branch and I saw a poisonous snake that had wound itself around it. I dropped both. We let everything wait until things dried out a bit.

Much later we reached our destination in the Hatin-Aga Gorge and made our camp. The night sky was starlit but the howling wind sounded with a terrifying roar. Nothing could protect you from it except your good reliable sleeping bag. I had a good bag but my companions did not. I remembered being in their place as a young post-grad when I covered myself with a wadded blanket over my padded coat for the night.

In the morning I got up, jogged to the brook to wash myself and to bring water for our tea as well as brush for the fire. For breakfast we ate a porcupine that Eduard had shot the previous night. The meat was delicious, no other words to describe it. The Turkmen usually do not eat porcupines, but here in the mountains away from his people, Geldi ate it with gusto. Some believed porcupine meat healed all diseases.

Natasha Burnasheva and I set out to find our object, the Rock-apple tree. I had traveled all over Central Asia and Trans-Caucasus and researched the majority of wild pomegranate areas. I understood that in some cases, for some reasons, a pomegranate settled in a rocky area may revert from a tall bush into a low shrub, or even into a spreading shrub, and occasionally, into a shrub crawling over the rocks. In the highest altitudes, on rock outcrops and in unfavorable conditions, some other woody plants revert into spreading bushes, among them, figs, Pallas buckthorn, Turkmenian maple, etc. This phenomenon is known as a phenotypic change—outwardly recognizable adaptations through the interaction of genes and environment. But you could never say in advance whether it was a purely phenotypic change or, quite the opposite, a genotypic change, a mutation, that determined hereditary transfer of a particular feature. To find out what we were dealing with, we had to transplant the shrubs into different environmental conditions, as I had been doing for many years with dwarf pomegranate plants.

We found the object of our quest and believed that the riddle of the Rock-apple tree would be solved. But the dwarf spreading plant we brought back to the station never survived planting. Such things sometimes happen. Thus it remained unknown how our strange apple would have behaved in the hot conditions of the Sumbar River Valley among the collection of other apple varieties. Would it remain a dwarf or grow tall?

Natasha and I never found the right time nor conditions for another expedition to collect more Rock-apple samples. We counted on future researchers doing what we couldn't finish—if there are any future researchers at Garrigala. As for Natasha Burnasheva, she was forced to retire. Sometimes, my wife and I phone her in Garrigala to wish her a happy New Year or to give her many happy returns on her birthday. Letters don't get delivered and we don't receive hers, but we can talk on the phone. The years of work, the research travels together, and now even the station whose name is no more—everything has changed beyond recognition. Only Natasha Burnasheva remains there.

### On The Road To Kughitang

In 1962, I made my first visit to that isolated mountainous country in southeastern Turkmenistan known as Kughitang. The second time, 15 years later, I came with Eghenmurad. The road to Kughitang was across cotton plantations on the right bank of the Amu-Darya River. It is the warmest region in the former Soviet Union and good for cotton, but because the area lay in what we called a "continental climate zone," characterized by hot summers and severe winters, subtropical fruit cultures, including pomegranates, had a subordinate role, growing only in private gardens.

The mountains and landscape made a deep impression on me. Here was the junction of Turkmenistan, Uzbekistan and Afghanistan. Kughitang was known for its caves, numerous semi-precious stones, native sulfur deposits, its deep lake, fossilized dinosaur footprints and distinct flora and vegetation. The area had no wild pomegranates, but people grew local varieties in their gardens. I remember well the Krk-Kiz Gorge on Kughitang-Uzbekistan territory. (Krk-Kiz means "forty young girls"—a motif popular in the east.) At Krk-Kiz, I was amazed to see abundant hybrid forms of local almond trees for the first time in my life. All morphological transitions between the varieties were represented, particularly noticeable in the pits and the fruit. Vavilov's expression fit here perfectly: this was a true "creation crucible."

In the caves, I was fascinated seeing stalactites and stalagmites for the first time in my life. The locals considered the cave a sacred place and hung little fabric strips to make their wishes come true. You saw this custom all over Central Asia. Near the Kughitang *kishlak*, the famous groves of small-fruited jujube still flourished on trees whose trunks were so thick that the four of us couldn't encircle one with our arms stretched wide. Fifteen years later, there were only pathetic remnants of that magnificent grove: the largest trees had died because they hadn't been irrigated sufficiently. In former times, a brook irrigated the grove, but an administrative boss had wanted its water for his own vegetable garden and diverted its course. Now they are creating a national park there. I am hopeful that things will improve.

# CHAPTER 27

# The Sacred Place of Shevlan

**Imagine a very late autumn.** The rainy season has come and gone and the weather is cold. Leaves on trees and bushes have turned yellow. At the station, roses and chrysanthemums are still blooming along with annual decorative plants. Pomegranates, late persimmons and olives have already been harvested. Porcupines certainly have partaken of pomegranates and found them to be to their liking.

Seven-and-a-half miles northeast of Garrigala, in the foothills of the Sunt Khasardag Ridge, there is also a little spring. There is an old *mazar* quite near to it. *Mazar* is a Central Asian name for a saint's grave over which a mausoleum has been built. This site was dedicated to Shivlan-baba *Mazar* who was a Sufi and Muslim philosopher, most probably alive in the seventh century A. D. The place was spelled "Sheblan" in the Soviet transcription.

Actually, it appeared that this devotional site had existed from an older, pre-Islamic time. I'm supported in this assumption by the stone lingam that was standing next to the *mazar*. The lingam was the phallic symbol used in the worship by Indian believers to represent the Hindu god Siva in his manifestation of virility, as a masculine source.

To get to the site, you took a road up several ridges that climbed higher and higher. The view that opened up onto the Sumbar River

Valley was magnificent. In spring, everything was turning green. In autumn there were bright spots of blooming yellow Sternbergia. Higher up, you saw several terraces where pomegranates and grapes were still growing.

Sunt Mountain dominated the site of Shevlan. At sunset, the usually dark-colored slopes appeared almost black. The silence over the mountains was stunning. Growing under the canopy of *derzhi* trees, you would find many Turkmen mandrakes. Local alcoholics illegally dug out mandrake plants though the plants were protected by law. They sold them for lots of money because there was a folk belief about that rare and sacred plant's mighty power of virility and potency.

In Shevlan, I enjoyed the hospitality of the pilgrims who came with their families from all over Turkmenistan and stayed in special houses. I had sometimes to stay overnight and they gave me shelter so that I could wake early the next morning and start my ascent of Sunt Mountain at sunrise while it was still cool. Ascent to its summit, 1,760 meters above sea level, was steep and hard. I can attest to it. I climbed it when I was young.

The site became crowded during Muslim holy days and sometimes pilgrims stayed a long time. Men slaughtered sheep, women cooked pilaf, mullahs prayed. In recent years, many mullahs had appeared. At Shevlan, there was work for everybody.

The vision of the valley and the serenity of the scene evoked in me a feeling of the eternal. It always reminded me of Isaac Levitan's painting, "Above the Eternal Peace and Calm." I was an irreligious Soviet scientist looking at a Muslim site and feeling as I imagined Isaak Levitan, a Russian Jew, had felt a century earlier above the Orthodox Christian cemetery that had inspired his profoundly reverent painting.

Interesting populations of wild pomegranates grew on the southern rocky slope of Sunt Mountain not far from Shevlan. It was there that I found the smallest pomegranate bush I ever came upon. It wasn't more than six inches high. By some miracle it grew in a narrow crack of a rock. It apparently had no resources to become larger, as the water that collected in the rock's cracked face did not suffice for it to grow to greater height. Of course I wanted to test my idea about why it was so small. I collected cuttings from that little bush. They rooted, and we planted the little plant on our special test site. But our eldest worker, Tore-Aga, had poor eyesight. He was digging on the site, did not notice it, and dug over our little plant just as it had begun doing well. What could you say to the lonely old laborer who had to do hard manual work to earn his bread in his advanced years? Alas, things like this happened more than once.

Levin in a pomegranate garden

# CHAPTER 28

# Destinies of Manuscripts

**One of Mikhail Bulgakov's characters remarked** that not all manuscripts burn. There are other ways they disappear.

In the 1920s, Nikolai Vavilov and his colleagues who studied fruit cultures prepared a multi-volume publication, *Pomology in the USSR*, describing all the varieties in the country. The book was ready to go but by then it was the 1930s and the Lysenkoists had come to rule Soviet agricultural science. Their interests were not in real science so the study never came into being. I particularly wanted to see one of Strebkova's monographs on pomegranates, but the manuscript was lost during the war. Again, this was the time of Lysenko, who unleashed his vicious pack of thugs on true scientists. Sometimes an honorable student preserved their professor's manuscript in hiding but thus far, Strebkova's precious work remains missing.

Decades passed. In the 1970s, interest in publishing that multi-volume work returned. I participated in a meeting with representatives from various agricultural stations in the USSR, and was assigned to prepare the pomegranate section. Everyone shared the enthusiasm and prepared articles, but the project came to naught.

For a quarter of a century, yet another manuscript on the best pomegranate varieties sat on the windowsill in a Soviet professor's office. The USSR will never publish our volume on subtropical fruit cultures. There no longer is a USSR.

Nikolai Vavilov used to say that every researcher should write at least one book. This wasn't what I did. I wrote numerous articles, mostly short articles. While I was getting ready for a longer work, the era, the epoch, changed. Vavilov was right when he said that one should hurry!

Our station was so far from cultural centers that the library exchange was our salvation and let us keep up on contemporary scientific work. Over the years, I have benefited from that exchange and received hundreds of books in Garrigala that I wouldn't have seen otherwise. We subscribed to all the major biological and agricultural magazines and could order new books ourselves.

My own library has suffered during my move, but I have taken some of my most important scientific books with me, and also treasured works of literature, including Bulgakov, Pushkin, Lermontov, Pasternak and Akhmatova.

Here in Israel, computers entered my life. It became easier to write, to communicate with friends and colleagues, and to obtain articles and other materials through the Internet and by e-mail. Such contacts and information have made my life full. They have also changed the destiny of my manuscripts.

At the end of the 1990s, Bill Feldman of the Boyce-Thompson Arboretum in Arizona offered his help in the publication of my manuscript on pomegranates. He took my Russian manuscript with him but money ran out before the translation was finished. I was convinced that some manuscripts are doomed by a bad fate hovering over them until quite recently, Feldman informed me that the translation of *Pomegranate* had been completed under the auspices of Richard Ashton, who grows pomegranates and apricots commercially at his Oak Creek Orchards in Texas. This came about after an American journalist, Barbara Baer, published an article on pomegranates at Garrigala and interested a number of other people in saving the collection.

When I was about to leave Turkmenistan, I'd had a conversation with Muhabbat Turdieva in Tashkent, Uzbekistan, who worked with the International Plant Genetic Resources Institute (IPGRI, headquarters in Rome, Italy) for Central Asia. She told me an American lady, Barbara Baer, was interested in our pomegranate collection and would donate money to publish a booklet bringing the plight of our station to world attention. I helped to write the text for the booklet. I was in Israel when Muhabbat sent me several copies. Barbara was quite persistent and found my address in Israel with the help of the agriculture department of the Israeli Embassy in Washington. She even came to Turkmenistan but was not able to visit Garrigala, only Ashgabat where she learned that I had already departed. In our e-mails that followed, she suggested that I write a popular book on the pomegranate. I postponed my scientific articles and sat down to write.

I have not met Barbara in person. We haven't spoken on the phone. My impression is that she is an energetic person who believes if there's something to be done, someone should do it. The "it" can be anything. In this case, pomegranates. And the "somebody" was herself. Isn't it interesting? Various individuals live at great distance from each other around the globe. They have their individual things to do. And one individual becomes the engine of something that he or she considers significant at the moment. That person involves others in some mutual activity. Barbara involved Margaret Hopstein, formerly of Tashkent, now a translator who lives in Seattle; Irina Slonimsky from Jerusalem; Jeff Moersfelder from U. C. Davis, California, who is a manager of our Turkmen pomegranate varieties in the USDA collections; and Richard Ashton who obtained my manuscript from Bill Feldman and brought it to Texas A & M Press. One more person, a specialist in the Israeli pomegranate nutra-pharmaceuticals, a director of a company in Haifa, Israel, also contacted me. Our connections have become a spontaneously formed International Pomegranate Club. Reasons to bring people together can always be found even in our troubled, turbulent and often violent world. As Bulat Okudjava's sang, "Let us hold each other's hands, friends…"

# CHAPTER 29

# Protection of the Species

**As our search for life on other celestial bodies** hasn't yet brought meaningful results, it's especially important that we preserve life on our planet in all its manifestations. If one has to worship anything, life alone deserves being worshipped at every level, especially now, when so many plants and animals are endangered. Every life form has the right to live, but many species are facing extinction.

The pomegranate is one of these endangered species. This is true along the northern border of pomegranate growth in the Soviet Union. I've observed losses with my own eyes. The genetic base of wild pomegranates is shrinking annually as the result of human pressure. Our direct influence has led to the loss of pomegranate plants, while indirect influence has altered the environment that supports the pomegranate.

The former Soviet Union had 170 Zapovedniks or national parks. Wild pomegranates within 15 parks in Georgia, Armenia, Azerbaijan, Turkmenistan, Uzbekistan, Tajikistan and Kirghizstan were protected. During my expeditions, I visited many of them and realized that we needed new and larger territories for the wild pomegranate populations. We need to do our utmost to preserve the cenospecies, all types of wild pomegranates within each ecosystem, so they are able to continue to exchange genes in the living wild gardens of their original growth.

## POMEGRANATE ROADS

IPGRI, the International Protection of Genetic Resources Institute, understands the problem. Several years ago an IPGRI expedition came to Garrigala and saw our collections, and also the wild pomegranate populations in the southwestern Kopet Dag. My impression was that the participants were impressed and were also concerned about the pomegranate population in the southwestern Kopet Dag. They understood we could not wait any longer to preserve the biodiversity of the pomegranate. Cultivated varieties of pomegranate are artifacts, creations of civilization and evidence of a very long, incredible history, the way a Rafael or a Brueghel painting or a Bach fugue is an artifact to protect and preserve for the future.

# CHAPTER 30

# As I Write by the Phoenician Sea

**I've probably left something unwritten.** Leonardo da Vinci did not finish the last hair in Mona Lisa's eyelashes. I've almost reached the end—enough is enough.

Convulsions of the twentieth century did not end with the last days of the past millennium. The beginning of the twenty–first century doesn't promise calm. There's much uncertainty. The world is not becoming homogenous. It probably never will. Multi-facetedness, polyvariability, is one of the principles of the universe. We need patience and optimism to believe that the world will reach some critical mass of democracy.

I write about transplantation into new environments. New soil, new impressions. *Omnia mea mecum porto.* Everything mine I carry with me.

I am not sorry about anything. It does not make sense to be sorry about the things that did not happen and are gone. I will not change. The very biological problems that were urgent for me before interest me now. Getting to their essence has no end and is my main life interest.

It's written that the loss of youth is the heaviest of trials, and that life becomes a sum of losses. I don't like to believe this. As always, I live thinking about tomorrow and new things. The desire to write is strong. Though it is more difficult to write about pomegranates being so far from my orchards, laboratories and the wild forests of my study, I nevertheless summarized my work in two scientific articles. I've recently been writing about succulents. My romance with succulents began years ago when I made experiments growing them on waterless slopes in Shihim-Dere Gorge.

The last three years in Israel have flown by. What will my tomorrow be like? I believe it will be the same. Though, as the saying goes, if you want to make the Almighty laugh, tell Him about your plans.

To be quite frank, I do not think I am the "Last of the Mohicans." I don't want the values of the past to disappear, to be annihilated, to fall into a Tartarean abyss. I believe that collapse of power, collapse of science in the former Soviet Union are only transitory, passing events. Speaking about Russia, Alexander Solzhenitsyn writes that Russia has lost the twentieth century. This is not true as far as Russian science. Nor is the statement that in much wisdom there is much grief, that he that increaseth knowledge increaseth sorrow, from Ecclesiastes. There is great satisfaction for the scientist who obtains new knowledge.

I want to believe that Russian science is recovering.

What is the future of the pomegranate? Like the rest of the world, the pomegranate is going global. There are enough mysteries about pomegranates to keep busy many generations of punicologists. Horticulturalists will create new varieties that weren't part of our ancient or modern gardens. Globalization of production and trade will determine the regions that are optimal for its cultivation.

Pomegranates are becoming important to chemists and pharmaceutical researchers because within the fruit's tough shell are found qualities to potentially relieve us of many illnesses and to restore our health in a world compromised by pollutants.

Pomegranates will find their deserved place, the way they once were in the world of western antiquity and the Orient. Maybe the ruby fruit will climb from eighteenth place in popularity of global fruits to perhaps something like tenth.

But the pomegranate is more than an edible fruit. It is a beautiful object, a decorative object. The pomegranate is an aesthetic image of our world, of its beauty, its unity, its uniqueness, its protectedness and unprotectedness.

I do not know whether pomegranates were fortunate to have me as their researcher but I certainly was very fortunate to have met the pomegranate, to be attuned to its energy, and to have my thoughts and deeds dedicated to it for many years.

I hope that the Planet of Pomegranates will always exist. I recommend to everybody, to you, my reader, to make a closer acquaintance.

March – August 2005/ Kfar Saba – Petah Tiqva, Israel

# ACKNOWLEDGMENTS

Cover painting, Daniela Naomi Molnar, 2006

Back cover photograph of Gregory Levin on the train to Daghestan, courtesy Collection of Gregory Levin

Back cover photograph of pomegranates, Marilyn Cannon

## Color Plates

1. "Twilight", Ann Getsinger

2. Silk Design Tapestry from France, 1734, Victoria and Albert Museum, London

3. "Ecstasy", Anitra Redlefsen

4. Punica granatum, illustration, Professor Otto W. Thome, *Flora of Germany, Osterreich and der Schweiz*, 1885

5. Bursting pomegranates, Marilyn Cannon

6. Black pomegranates, Gregory Levin

7. Flowers and buds, Anitra Redlefsen

8. Turkmen, Gregory Levin

9. Three mature pomegranates, Marilyn Cannon

10. Pomegranate tasting at UC Davis, Marilyn Cannon

11. "Pomegranate III", Celia Gilbert

## Black and White Photographs

Three pomegranates, Anitra Redlefsen

Pomegranate Crown, original drawing, Allis Teegarden, 2006

All other photographs courtesy Collection of Gregory M. Levin

# ORDER FORM

Fax or mail orders with check to:

Floreant Press
6195 Anderson Road
Forestville, CA 95436
Fax/Telephone: 707.887.7868

Please send me _____ copies of
*Pomegranate Roads* at $18.00 each.

Please include $3.00 shipping and handling for the first copy
and $1.00 for each additional copy.

*Ship to:*

Name _____

Street _____

City _____

State _____ Zip _____

Phone or email (in case we need to contact you about order)

_____

Order directly online:
www.floreantpress.com